Leading with Safety

Leading with Safety

Thomas R. Krause, Ph.D.
Chairman of the Board
BST

WILEY-
INTERSCIENCE

A JOHN WILEY & SONS, INC., PUBLICATION

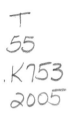

T
55
.K753
2005

Published by John Wiley & Sons, Inc., Hoboken, New Jersey.
Published simultaneously in Canada.

For general information on our other products and services or for technical support, please contact our
Customer Care Department within the United States at (800) 762-2974, outside the United States at
(317) 572-3993 or fax (317) 572-4002.

Wiley also publishes its books in a variety of electronic formats. Some content that appears in print may
not be available in electronic format. For information about Wiley products, visit our web site at
www.wiley.com.

Library of Congress Cataloging-in-Publication Data is available.

ISBN-13 978-0-471-49425-6
ISBN-10 0-471-49425-9

Printed in the United States of America.

10 9 8 7 6 5 4 3 2 1

Leading with Safety

Contents

Contents

Foreword

John L. Henshaw

In my 30 years of professional practice I have seen many organizations struggle in trying to find the right mix of corporate statements, organizational structure, policies, practices and training techniques for improving workplace safety and health. While these are critical steps in reducing hazards and reducing injuries and illnesses, they do not provide the complete answer to achieving superior performance. The key to moving from the current status quo to a higher performance level just might be within the covers of this text. *Leading with Safety* provides some critical insight into understanding the relationship between the behaviors of leaders and organizational culture and making the move to a truly high-performance organization — an organization that excels in workplace safety and health as well as other business performance indicators such as quality, productivity and profitability.

Having served as Assistant Secretary of Labor for Occupational Safety and Health (OSHA), I saw four types of organizations: those that exercise blatant disregard or indifference to worker safety and health; those that are just plain ignorant of workplace hazards and relevant laws and regulations; those that are committed to worker safety and health and trying to improve their performance but have fallen short of expectations, and; organizations that have the right stuff and have achieved true superior performance.

The first group, those that blatantly disregard the law, are and should be the focus of OSHA's enforcement efforts. They represent a small percentage of workplaces in this country in my view and should remain on the Agency's enforcement priority list until there is a change in organizational culture, behaviors, management systems and performance.

The second group includes those organizations that just don't know or may not have the resources to fully understand what is required of them in respect to worker safety and health. In addition to enforcement, OSHA uses their extensive outreach, training and compliance assistance efforts to help them reach compliance and achieve greater worker protection. Many of these organizations are small- to medium-sized workplaces and when given the opportunity to understand and realize the value of worker safety and health will continue to try to improve performance.

The third group includes organizations that recognize the value of worker safety and health and are trying to make improvements but have failed to make the step change. They often say the right things, have good management commitment and have implemented sound management systems and practices but superior performance still eludes them.

The last group are those organizations that not only say the right things, they also practice the right things, behave the right way, have the right systems in place, engage all their employees and surpass most of their counterparts in safety and health performance. Many of these workplaces are in OSHA's Voluntary Protection Program (VPP) and more often than not they outperform most of their peers in other critical general business measures such as environmental compliance, productivity, quality, morale and profitability.

Why do organizations such as those in the last group excel? The answer in my view is in behaviors, in particular the behaviors of management. This text provides some very useful insights into behaviors and how an organization can begin to address the psychological side of safety and health. Successful organizations have been able to master the elements of organizational safety leadership and engage workers at every level in the process of creating an injury-free workplace. They understand the crucial aspects of human performance and fully realize that to continuously improve performance and achieve superior results the organizational culture must change — and for the culture to change behaviors must change. What sets them apart from the rest is that they address the behaviors of all employees, including the CEO and senior managers through to the front-line supervisors and hourly workers. They strive for behaviors that support superior performance not only in safety and health but in all aspects of the business.

The vast majority of workplaces and the vast majority of workers in this country fall in the second and third groups. For them this book will help explain the key elements of safety leadership and commitment and dispel myths about applied behavioral analysis, which is an essential tool for creating supportive behaviors and the desired culture. More importantly, they will learn how personal values, leadership style and what leaders do and do not do, impact the culture of the organization and ultimately its overall performance.

Clearly, the essential key to achieving superior safety and health performance is engaging employees at all levels of the organization. This book describes and gives practical applications and considerations on how engagement can be enhanced in an organization and the role the front-line employees, supervisors and senior leaders play in creating an injury- and illness-free workplace. It also provides some useful tools in coaching senior executives who must be effective leaders. Of course "what gets measured-gets done" and this text outlines discrete elements of an organizational culture and safety climate that can be measured and that correlate well with safety and health performance.

As engineers, scientists and business managers, we often see our operations as a series of transactions. We falsely believe that if we document and communicate our policies and procedures, establish good management systems and train, train and train all our workers, we can operate our process according to plan and achieve superior performance. But because workplaces are dynamic and dependent upon the interaction between technology and people, human performance, as described in this text, is the missing link to achieving superior results. Any organization interested in establishing an incident-free and injury- and illness-free workplace can benefit from taking stock in the ideas, data and systems described in this book. It could open the door to new ideas and identify powerful tools to take an organization closer to that ultimate target.

• • • • •

Preface

We have written this book to summarize the things we have learned in the last five or six years about the relationship between safety leadership, organizational culture, and organizational performance, and their implications for safety performance improvement.

Starting in 1994, we began an outcome study designed to quantify the results that BST client companies were getting with the methods we had taught them. This research, which was later published in a peer-reviewed journal[1], tracked seventy-three individual projects aimed at safety improvement. The study found that *on average* each of the seventy-three organizations reduced the frequency of incidents by about 55% over a five-year period.

But what really caught our attention was something we had not even intended to study. We were surprised by the *huge variability in incident frequency reduction across the seventy-three organizations.* Some got results quickly, almost immediately, and maintained them over the five-year period. Ten companies reduced incident frequency by an average of 80% after two years. Others took several years to start to improve, but eventually got the results they wanted. However, some got virtually no results while a few got worse. Perplexed at this level of variation, we became fascinated with understanding the factors that accounted for it.

We designed a study to answer the question: *Why do some organizations do well with safety improvement initiatives while others do poorly or fail?* The research method on which we settled was an "extreme groups" design. We

---

[1] Krause, T.R., K.J. Seymour, and K.C.M. Sloat. 1999. "Long-Term Evaluation of a Behavior-Based Method for Improving Safety Performance: A Meta-Analysis of 73 Interrupted Time Series Replications." *Safety Science.* 32: 1-18.

identified two clusters of organizations that represented both ends of the improvement continuum — the best and the worst. Our team studied these organizations carefully with site visits, surveys, and interviews, and analyzed their data. The objective was to isolate factors most strongly related to the outcomes we had found: success and failure.

The results of the study published in 1998[2] comprised a set of critical success factors found to be common in the successful organizations and lacking in those that failed. But two findings stood out from the rest, one formal and explicit and the other informal and less well defined:

1) The most important factor in predicting success of safety improvement initiatives was the quality of leadership they were given and the organizational culture that resulted. This may not be surprising to anyone familiar with successfully implementing performance improvement initiatives generally. But it was interesting to us because the initiatives we were studying were employee-driven. It was unclear to many leaders in these organizations what role they should play in employee-driven initiatives — or whether they should play any role at all. That was ironic given our finding that the kind of leadership they provided was pivotal to the company's safety success. Our research also showed that effective safety leaders demonstrated particular characteristics, engaged in specific behaviors, and tended to create a certain kind of organizational culture.

2) Companies highly successful in safety were also successful in operational performance generally. Again, this may not be startling to the experienced observer of organizational performance. But what are the implications of this fact? Why does excellence in safety performance go along with operational excellence? We became intrigued with the question, "What is the relationship between excellence in safety performance and overall operational excellence?"

Consequently, we became very interested in safety leadership and its relation to organizational functioning. As we began to study what great safety leaders do, how they lead, and what makes them effective, a picture began to emerge that we thought had serious implications for helping companies

---

[2] Hidley, J.H. 1998. "Critical Success Factors for Behavior-Based Safety." *Professional Safety.* 43: 30-34.

improve in safety and in organizational functioning generally. For the past three years, we have worked on developing methods to support safety leadership and on specifying the mechanisms that connect safety leadership to high levels of organizational functioning. Many ideas outlined in this book represent refinements and improvements to our previous work. Some thought processes are entirely new, and many orientations have changed as we have learned from experience.

This book is about those methods and related mechanisms. It is written for *safety leaders at any level in the organization.* By "safety leader" we mean any employee who has an influence on safety in the organization. It is unusual to write a book intended for such a wide variety of people throughout an organization. We are taking this approach because we have learned from experience that safety considerations are similar across employee levels and that safety leaders are found at all organizational levels. Safety happens at the floor level of the organization; it is ultimately about the worker interacting with the technology. But each leader has a unique role to play in assuring that the worker is protected from exposure to hazards.

We sponsor an annual conference for our clients working on safety improvement. At our 2005 event, about 3,000 people from 420 organizations attended. Every organizational level was represented and participated. Break-out sessions ranged from union concerns to senior executive leadership skills training. Conversations with CEOs differ little from those with first-line supervisors and informal leaders from the front-line employee group. They are all about how to create the right environment, or safety climate, that will lead to a culture in which injuries and incidents are unacceptable: *an injury-free culture.* Of course, each level has a different contribution to make in creating this environment. But the interesting aspect organizationally is that *employees at all levels share the common ground of getting safety right.*

· · · · ·

Acknowledgments

This book is the result of an ongoing collaboration among three groups of people: BST consultants, client organizations and technical resources. I would like to thank some of those who have made this effort one of the most fulfilling in my professional life.

Consultants are always indebted to their clients. This work was made possible by the individual organizational leaders who saw that safety leadership and organizational culture could and should be improved, and put resources on the line to do it. We have had the privilege of working with many great senior safety leaders in outstanding organizations. There are too many to name them all here, but I would like to mention several that stand out: Sean O'Keefe at NASA, Paul O'Neill when he was at Alcoa, Jim Dietz at PotashCorp, Andrew Stephens at Petro-Canada, Paul Anderson at Duke Energy, and Tom Weekley of the United Auto Workers.

Second, my colleagues at BST, some of whom contributed seminal ideas that stimulated my thought process and led to the overarching ideas in this book. The entire BST consulting team guided the development of these ideas, providing feedback and fine-tuning as ideas turned to action. Some wrote entire chapters, others collaborated with me on the writing. Among the collaborators were John Hidley (Chapters 2, 9, and 10), Kim Sloat (Chapters 4 and 8, including much of the underlying research), and Rebecca Timmins (Chapter 6). Credit for writing entire chapters goes to Jim Huggett (Chapter 7); Don Groover, Jim Spigener, Rebecca Timmins, and Ken Jones (sections of Chapter 11); and Scott Stricoff (Chapter 12). Kristen Bell organized and conducted much of the research, with the support of Dave Hoffman at the University of North Carolina. Credit for seeing the need for a Safety Leadership Model in the first place goes to Pat Smith, who constructed the first of several iterations leading to the final version in this book. Pat also saw the need for an Organizational Safety Model that would convey the "big picture," and contributed to its development. And Rebecca Nigel and Bonnie Irving did the painstaking job of editing that pulled it all together, while Marty Mellein designed graphics and layout.

Finally, on a personal level, I would like to thank my wife, Cathryn, and our four children Elizabeth, Christel, Leslie, and Jared, for their support and encouragement during the writing of this book. They put up with me during the stressful periods and for that I am grateful. I would also like to thank my friend and coach Phil Fedewa, for his thoughtful insights on how to get things done effectively while maintaining order and balance across the wider range of personal objectives, and Leslie Hidley for her support and encouragement about the writing itself.

• • • • •

Safety as a metaphor for organizational excellence

You are the new CEO of a large manufacturing organization. Your organization is well established and has valuable assets, but profits are level, employee relations are mixed, morale is relatively low, and market conditions are not particularly favorable. You realize that to improve performance you will need to *engage employees at all levels of the organization in new and meaningful ways.* You know that this engagement can make all the difference in turning things around.

Easy to say. But it doesn't take long to discover that serious issues stand between your improvement goals and the real world in which your employees live. The organization is not unified, and leadership lacks credibility. Supervision is often weak and is perceived as arbitrary and even unfair. Recent surveys show that employees don't think leadership is listening. When they are asked how they see leaders, they answer, "They don't care." Recent initiatives to improve operations have been mediocre, largely because employees don't buy into the new message. Safety performance is about average for your industry and has been for the last several years.

How do you approach the situation? What are the strategic considerations that guide your thinking? Your consultants tell you that *you need to change the culture of the organization,* and your executive team agrees. But you aren't naïve enough to think you can manage culture change as if it were a production schedule or an engineering analysis. And your background is engineering, finance, or marketing, not anthropology, sociology, or psychology. You know from experience that despite all the fancy talk from culture consultants, many leaders have taken on the culture issue and been frustrated with either slow tangible results, or none at all. And when you think about it, if culture is essentially made up of values, how is it reasonable to think that *any leader can actually change the values of employees?* Getting together a strategic plan in which you have a high degree of confidence will not be easy.

Yet, this is the situation *leaders* face in organizations all over the world. Whether first-line supervisors, site managers, or CEOs, the set of issues described above is more common than unusual. *The premise of this book is that these issues can be addressed efficiently, effectively, and reliably by using a method that many will find unusual and even surprising: leading with safety.* That means getting the organization mobilized to do something it already values: preventing employees from being injured or killed, on or off the job. As consultants to organizations for the past twenty years, we have seen and helped many companies develop and implement this strategy, with remarkable results. Yes, safety improves and fewer injuries, illnesses, and fatalities happen, but perhaps even more interestingly, when safety improves, a whole set of other performance indicators improves with it. As well, the culture changes in highly significant ways. And culture change is clearly a leadership issue.

There are several mechanisms that explain culture change. One is the principle of reciprocity discussed in detail in Chapter 4. Another is that employees who know they are cared about are receptive to engagement. Still another is that operational excellence parameters are highly correlated. The best locations are not just best in quality or production; they tend to be best across the board. The important point is that it matters where you start the improvement process. And safety is an ideal place to start.

The most well known exemplar for this model is Paul O'Neill, who as Alcoa's CEO faced a situation very similar to the one described above in 1987. He realized that to transform the organizational culture at the aluminum giant, he would have to find a way to engage employees at all levels. And this engagement had to occur in the context of cultural unity; employees had to believe the organization really cared about what happened to them. So, O'Neill did something extraordinary: he made safety outcomes the primary indicator of senior leadership's performance. Over time, he transformed the culture of Alcoa, which became a world leader in safety performance, and set the stage for the kind of *organizational functioning* it needed to grow and prosper.

In 1987, Alcoa's lost-time incident frequency rate was 1.86. In 2002, it was .12.[1] In 1987, net income was $264 million on sales of $4.6 billion, with 35,700 employees and a market cap of $2.9 billion. In 2000, when O'Neill retired, profits stood at $1.5 billion on sales of $22.9 billion, with 140,000 employees. Market cap was $29.9 billion.[2]

What it Means to "Lead with Safety"

We call this approach to organizational change "leading with safety." The phrase also means several other things. Leading with safety means that safety performance excellence "leads" the organization to other kinds of performance excellence. It also stresses the importance of *leadership* to safety excellence. And it means that leaders who lead with safety establish themselves "as leaders" in a different, valuable, and interesting way.

How This Book is Organized

Some great leaders champion safety improvement as a way to create cultural unity, improve organizational functioning, and enhance operational excellence. But in many organizations, they are the exception. We must start the improvement process with the leaders we have, building on their strengths and supporting their opportunities to improve. How does the change agent begin to influence the rest of the organization to *lead with safety*?

[1] Smith, S. October 2002. "America's Safest Companies: 17 Award Winners Share Best Practices." *Occupational Hazards*. 47-62.

[2] Arndt, M. February 2001. "How O'Neill Got Alcoa Shining." *Business Week*. 39.

Throughout the book, we will provide answers to this question from different perspectives. In Chapter 1 we will examine core concepts leaders need to understand to be effective safety leaders, including an Organizational Safety Model that provides a big-picture perspective. The next three chapters consider a Safety Leadership Model. Many organizational leaders have asked us, "What does it take to be a great safety leader?" We have studied this question carefully after working with many leading organizations and the Safety Leadership Model is the result of this study and our experience. Chapter 2 looks at the individual safety leader's personality and values, emotional commitment, and leadership style. It addresses questions from the standpoint of the individual: what personal attributes are required to be an effective safety leader? Chapter 3 reviews best practices in leadership behavior: what do effective safety leaders do in terms of specific behaviors? And Chapter 4 provides an analysis of the core elements that comprise organizational culture and safety climate.

Chapters 5 through 9 deal with specific tools that assist the change process. Chapter 5 is given to understanding applied behavior analysis, a powerful but easily misunderstood methodology, and Chapter 6 to the fascinating field of cognitive bias. Chapter 7 is about coaching senior executives to be effective safety leaders. Chapter 8 deals with the unique role of the supervisor, and Chapter 9 with a specific methodology for improving worker safety at the site level.

We then turn to applications. Chapter 10 is given to strategies for the design of interventions using the methods in this book, and Chapter 11 covers case history accounts of how organizations have used the leading-with-safety approach.

Finally, in Chapter 12 we review the safety culture change process undertaken by NASA after the Columbia tragedy.

Since this book is designed to close the gap between theory and practice, each concept presented in the Safety Leadership Model has hands-on methods designed to put the concepts to use. The CD in the back of the book will facilitate that process.

Section 1

The Organizational Safety Model

We begin with a "big picture" model showing the core elements
of organizational safety performance

Chapter 1

The Organizational Safety Model: Understanding the Big Picture

- The Organizational Safety Model:
  How does safety leadership assure improvement?

- The primary importance of the Working Interface

- Understanding the relationship of exposure events to injury events

- The necessity of leading indicators

- Enabling safety systems

- Sustaining safety systems

- Leadership creates organizational culture and safety climate

- What motivates leaders to improve safety?

- Influencing the behavior of safety leaders

- Sustaining organizational change: Two critical elements

Note: Throughout this book, the phrase "safety leader" means any person who influences others in the organization regarding safety. This includes the senior-most leader in the organization, all managers, labor representatives, and workers. We also use the terms "safety" and "EHS" (Environment Health and Safety) interchangeably. Our clients use a variety of acronyms to describe these areas of functioning: SHE, HSE, etc. We see the three areas as interrelated and the methods that affect one as appropriate for each of the others.

Effective safety leaders have a solid understanding of a set of core concepts. For those with operations backgrounds, these concepts have been learned from experience and are now intuitive. Leaders from other areas will find these essential concepts easy to grasp, but challenging to execute effectively. Understanding these core concepts is a pre-condition for creating excellence in the leadership of safety performance and provides a systematic way to think about the design of intervention strategies for improvement.

The Organizational Safety Model: How Does Safety Leadership Assure Improvement?

Leaders need to understand, in concrete terms, how safety leadership assures improvement. We know leadership is important to safety excellence generally, but what are the specific mechanisms that connect the leader to safety performance improvement? As a safety leader, I need to know what specific mechanisms produce results. It is not sufficient to say that "management needs to support the safety effort." Of course management support is needed, but we need to be much more specific about what is expected of the safety leader to assure his or her success. What are the common threads that run through safety improvement mechanisms, and what behaviors does the leader engage in to assure them? How do the things I do and say, or fail to do and say, influence the work configuration and the way work is done?

The effective safety leader understands the "big picture" shown in the Organizational Safety Model, Figure 1-1. It addresses how leadership and

Leadership

Safety Enabling Systems

Hazard Recognition and Mitigation

Skills, Knowledge and Training

Regulations and Procedures

Safety Improvement Mechanisms

Organizational Culture

Organizational Sustaining Systems

Accountability

Selection & Development

Organizational Structure

Performance Management

Employee Engagement

Management Systems

Equipment

Working Interface

Worker   Facilities

Procedures

LEADERSHIP
Seeing the right things to do to reach objectives and motivating the team to do them effectively. Safety leadership is exercised by decision-making, which is related to the beliefs of the leader and demonstrated by his or her behavior.

SAFETY ENABLING SYSTEMS
The set of mechanisms that enable safety in the Working Interface. Different organizations classify these in different ways, but usually include basic safety mechanisms: hazard recognition and mitigation, training, regulations, procedures, policies, and safety improvement mechanisms.

ORGANIZATIONAL CULTURE
The driving values of the organization. "The way we do things around here." The unstated assumptions about how things are done. Distinguishable from safety climate, which is the emphasis perceived to be given to safety by the organization's leaders.

ORGANIZATIONAL
SUSTAINING SYSTEMS
The set of systems that sustains enabling safety systems and assures their effectiveness. This includes selection and development of people, performance management, organizational structure, employee engagement, and other management systems.

WORKING INTERFACE
The interaction of equipment, facilities, procedures, and the worker. A combination of these factors creates or eliminates exposure to hazards.

Figure 1-1. The Organizational Safety Model.

culture influence the organization's safety enabling systems and organizational sustaining systems to reduce exposure to hazards in the Working Interface.

The Primary Importance of the Working Interface

The Working Interface is the configuration of equipment, facilities, systems, and behaviors that define the interaction of the worker with the technology. Hazards exist in this configuration. Safety excellence is directly related to how effective the organization is at controlling exposure to hazards in the Working Interface. Essentially, safety concerns how workers interact with the organization's technology. Each element in the Organizational Safety Model plays a critical role in controlling exposure to hazards.

Enabling and sustaining systems are designed to reduce and eliminate exposure to these hazards, either by eliminating the hazards themselves (e.g., through modified production processes) or by introducing hazard control measures (e.g. guarding or venting systems).

We have intentionally avoided saying what proportion of incidents comes from what type of exposure in the Working Interface. Many in the safety community believe a high percentage of incidents, perhaps 80-90%, result from behavioral causes, while the remainder relate to equipment and facilities. We made this statement in our first book published in 1990. However, we now recognize that this dichotomy of causes, while ingrained in our culture generally and in large parts of the safety community, is not useful, and in fact can be harmful.

There are several reasons for that. First, the dichotomy is not representative of what actually happens to cause injuries. The equipment doesn't simply malfunction, independently of how it has been designed and maintained, and the worker doesn't simply behave unsafely, independently of the system configuration. Rather, the worker interacts with the technology, and the interface that results comprises a system. Multiple variables influence the system: the quality of design, appropriateness of training, influence of culture and climate, and the quality of leadership.

Second, the dichotomy tends to encourage blaming. If the purpose of understanding what causes injuries is to establish fault, it is useful to have neat (although inadequate) categories like "worker behavior" or "equipment and facilities" as sources of the injury. This is a natural reaction and one seen regularly in the popular press — "the accident was the result of operator error." But it is often counter-productive because it leads to blame. And blaming is always a mistake.

The useful question is not "Who was at fault?" but rather "How can this injury, and others like it, be prevented in the future?" If we fail to realize this we fall into the trap of arguing over fault, and the process of understanding why the injury occurred becomes biased by various points of view that want the outcome of not being blamed! Anyone familiar with the incident investigation process in a weak organizational culture has seen how destructive this process can be. Incident investigation committees can waste time, make poor recommendations, and undermine the safety climate at the facility, whether by calling things "operator error; instruct the operator to act differently" or by seeing everything as facility-related or the fault of management.

High-functioning safety organizations have gone beyond the entanglements of blaming and recognize that getting safety right means designing and influencing systems that reduce and eliminate exposure.

Of course, not all exposure is equal in terms of the potential for serious injury it represents.[1] Some exposures to hazards will result in more serious incidents, some in less serious ones. [2]

Understanding the Relationship of Exposure Events to Injury Events

Leadership must understand *the relationship of* **exposure** *events to* **injury** *events*. H.W. Heinrich first described this relationship in 1959[3]. It has been used in most standard safety texts since. It is expressed as the familiar "safety

---

[1] Manuele, F.A. February 2004. "Injury Ratios". *Professional Safety*. 26-30.

[2] Defining the precise relationship needs to be done on a case-by-case basis and will vary by type of industry, and a number of other factors.

[3] Heinrich, H.W. 1959. *Industrial Accident Prevention, 4th Ed.* New York: McGraw Hill.

triangle" showing fatalities at the top and less serious injuries further down in the base. Recently this concept has been criticized[4,5] because it does not account for the fact that some hazards are much more likely than others to cause serious injuries, and so there are dramatic differences in the ratio of major to minor injuries depending on the specific hazard in question. Critics argue that when the focus is on frequency alone, emphasis can be mistakenly placed on those hazard types associated with many minor injuries — rather than focusing on the smaller number of minor injuries related to hazard types more likely to be predictive of the risk of major injuries. This criticism is valid in our view. It would be simplistic to assume that all exposure events are equal in the severity they represent or would warrant the same types of intervention. However, *it would also be a serious mistake to forget that more frequent low-severity events may be indicators of the potential for high-severity events.*

Actually, this concept is very useful, in safety as well as in other applications. It can be used to understand a wide variety of "unwanted events" — outcomes recognized as undesirable yet difficult to predict and control. The critics of this principle often fail to realize its significance and implications. Examples range from catastrophic events to crimes, college dropout rates to medical errors, just to mention a few. The underlying principle states that many small or less severe events precede a single large or serious one. Those smaller or less severe events may be similar in type but lower in severity (e.g., small leaks vs. a large one) or may be precursors on a chain of events leading to the major event (e.g., not blanking a line, leading to a major fire).

This principle has two significant implications: a) when a single serious event occurs, it can be inferred with high probability that many related smaller events have occurred previously; and b) to prevent workplace incidents, small events and their precursors must be taken as seriously as large ones. Each time a worker is exposed to a hazard, that exposure represents an important risk, whether the actual result is no harm, a fatality, or another

[4] Kriebel, D. 1982. Occupational Injuries: Factors Associated with Frequency & Severity. *International Archive of Occupational & Environmental Health* 50: 209-218.

[5] Manuele, F.A. 2002. *Heinrich Revisited: Truisms or Myths*. Itasca: NSC Press.

type of catastrophe. This fact has very important implications for the practical day-to-day things we do in organizations to improve safety. It means that an injury-free culture is not simply one that doesn't tolerate incidents. *An injury-free culture is one that doesn't tolerate exposure to hazards.*

This principle provides fresh insight when applied to performance issues outside of safety. As an example, let's look at a student living in a college dormitory who is found to be using heroin. The college administration wants to know if this is an isolated event, or whether other students are also using the drug. The "exposure to event" principle suggests that the probability of a single serious event occurring in isolation is very low and that if one serious event has occurred, it is *highly likely* that many other less serious events of the same kind have happened previously. In this case, one student using heroin indicates that other students on campus are probably doing the same.

Similarly, if a single fatality occurs in a manufacturing environment, *it is highly likely that its causal roots — the particular exposures to hazards that caused it— occur frequently in the Working Interface.* (This is not to say it is always present. If the Working Interface is free of this exposure the great majority of the time, but it is present even five percent of the time, we can expect injuries to follow.) Leaders are often slow to recognize this statistical fact. It's easier to think that the serious event represents a "fluke." As the statistics professor used to say, "Rare events happen, *rarely.*" Most of the time, the outcome is unusual, but the events leading up to it are common.

Let's explore an example. We observe a car crash on the road to the office. For some unknown reason, the driver swerved into oncoming traffic. He was not wearing a seat belt and was thrown from the car and killed. What is the probability that the driver usually wore a seat belt, but didn't do so that day? It could happen, but the likelihood is very low. It is much more likely that the driver frequently did not wear a seat belt, and the day of the accident was no exception.

Similarly, a worker falls to his death from an elevated surface and the incident investigation finds that he was working on inadequate scaffolding and not wearing fall protection. What is the likelihood that inadequate scaffolding and failure to wear fall protection are rare events within that facility? It is possible, but the likelihood is low. It is more likely that these hazards occur with regularity and have previously produced smaller events like near misses or minor injuries. *Further, this means that the leadership has tacitly accepted these events.*

To review, an exposure event is the exposure of a worker to a hazard within the Working Interface. It includes exposures related to facilities, equipment, actions, and most importantly, their interactions.

The same number of exposure events in a given time period will lead to a different number of incidents in the same time period. This is a statistical fact stemming from random variability. A given exposure today has a different result than it will tomorrow, simply by chance. If this basic statistical relationship is not understood, leadership will inevitably overreact to incident data. A few months will go by in which injury frequency is unusually low, and leaders will conclude that safety is actually improving, when in fact it may or may not be. Or in a period of a few months, a "rash of injuries" will occur, after which leaders say that safety has deteriorated when in fact exposure may have been reduced. None of this is to say that safety is ultimately a matter of luck, but it does say that *incident frequency is subject to random variability.* Effective safety leaders need to understand these relationships.

The Necessity of Leading Indicators

To understand the relationship of exposure events to injuries requires reference to leading indicators. Leading indicators are measures of variables that can be shown to have a statistically valid, predictive relationship to injury frequency. When viewed in relation to lagging indicators (for example, the number of injury events divided by hours worked), leading indicators allow organizations to take proactive measures that prevent injuries. Figure 1-2 shows some leading indicators.

Perhaps the most interesting aspect of all this is that most safety professionals already understand it, while *safety leaders,* including executives who make important safety-related decisions, frequently don't.

Leading	Lagging
· Working Interface assessment	· Environmental events
· Hazard assessments	· Injury rate
· Working observations of exposures (e.g. % safe)	· Illnesses
· Safety participation rates	· Workers' compensation costs
· Near misses	· Lost work days

Figure 1-2. Leading and Lagging Indicators.

Enabling Safety Systems

Enabling safety systems combine to reduce and eliminate exposure to hazards. These are the basic safety systems or programs that assure adequate safety functioning. Most large organizations have them in place and audit them regularly. The effective safety leader knows what these systems are, how they are audited, and how effective they are.

Interestingly, most organizations have discovered that two sites can have practically identical audit scores of site-level enabling systems, identical or near-identical technology, and similar workforces, and yet report *very different incident frequency rates*[6]. Enabling systems are necessary but not sufficient for excellent safety performance. As we will see, there is a great deal more to safety success than enabling systems.

---

[6] Petersen, D. April 1998. "The Four Cs of Safety: Culture, Competency, Consequences & Continuous." *Professional Safety*. 32-34.

Sustaining Safety Systems

An organization's sustaining safety systems are those that sustain safety performance excellence. Most organizations have these systems, but variation in their quality is much larger across organizations than with enabling systems. Perhaps more importantly, most organizations fail to appreciate the relationship of these systems to the Working Interface, to safety enabling systems, and to safety performance outcomes.

Is safety leadership a criterion for jobs that are central to safety outcomes? Is there a process to develop safety leaders? Is the structure of the organization such that safety is given adequate emphasis? Does the performance management system meaningfully address safety leadership issues (not just through lagging indicators)? Are there mechanisms to assure employee engagement in safety? Is there a systematic way of holding leaders accountable for safety processes and outcomes?

Leadership and culture determine how well these systems work. But saying that is the easy part. Leadership has the task of continuously improving both enabling and sustaining systems. But the workplace is complex: technology changes, organizations change, operations pressures exist, cultural factors may not be ideal, and so on. Given the real world, how does the safety leader assure safety improvement?

Leadership Creates Organizational Culture and Safety Climate

Ultimately, the safety improvement objective is to create a positive safety climate and a culture in which safety is a driving value. But if we look realistically at many organizations, we often see issues such as low trust, poor communication, and mixed management credibility. Many leaders are failing to address hazards and front-line employees are often not engaged. How do we change all that? How do we create a culture in which safety really is a driving value?

The change process starts with leadership itself. A core group of leaders who have influence over the organization needs to get aligned on what they really value, and what principles represent those values. Then they need to know what behaviors of theirs are necessary to convey to the organization that they are serious about change and to stimulate the right behaviors among other leaders. There must be consistency across leadership on saying and doing the right things: making the right decisions, communicating the right information, and articulating the right vision.

Culture changes slowly, but it is changing all the time. Leaders are always changing the culture each time they make a decision, leave an issue hanging, take a stand, or address an issue. The change process is about directing and accelerating the natural change that is already happening. That it takes time to change a culture is both good and bad. It means that a weak or ineffective culture will take a long time to change, but at the same time, it means that when safety becomes a driving value in the organizational culture, that value will endure.

The strength of site-level enabling and sustaining systems is not sufficient to predict variability in performance. To understand the reasons for variation in the frequency rates of incidents, safety leaders also need an understanding of *organizational culture and safety climate*. If two locations have similarly well-developed enabling and sustaining systems, similar technology and workforce, but different incident frequency rate levels, the difference between them will likely be found in their cultures.

Organizational culture is the shared values and beliefs that drive behavior in an organization — commonly described as "the way we do things around here." The concept of culture is widely understood, but the relationship between organizational culture and the safety climate, and their role in safety excellence, is not.

We discuss culture and climate in depth in Chapter 4, but for now, we will provide a brief overview. Where organizational culture involves unstated assumptions that govern how we do things around here, safety climate refers to prevailing influences on a particular area of functioning (safety in our case) at a particular time. Thus, organizational culture is deeply embedded and long-term; it takes longer to change and influences organizational performance across many areas of functioning. Safety climate, on the other hand, changes faster and more immediately reflects the attention of leadership. Think of organizational culture as background influence on the organization, while safety climate is foreground. Climate changes faster than culture.

The safety leader must do more than influence site level improvement elements. He must also take on the task of influencing the organization's culture and safety climate. Further, *site level safety improvement elements are managed, but organizational culture and safety climate are led.* Chapter 3 addresses the difference we intend between management and leadership, as well as specific best practices in safety leadership. In short, management has to do with *what things get done*, and leadership has to do with *how they get done.*

Ultimately, the safety leader's job is to contribute to and support mechanisms that reduce and eliminate exposure to hazards.

What Motivates Leaders to Improve Safety?

Based on our experience in working with leaders ranging from first-line supervisors to CEOs, three primary motives drive safety improvement: feeling compassion, building a performance platform based on cultural unity, and contributing to profitability.

When we first began working with senior executives many years ago, some on our team were surprised to find that while leaders vary, *the predominant motive driving senior leaders to improve safety is human compassion.* We find this holds true for safety leaders generally, whatever their level in the organization. The leader who works to improve safety is usually doing so out of a deep sense of integrity, a grounding in ethical principles, a feeling that it is the right thing to do.

This kind of motivation differs fundamentally from other business motives. The leader's motivation to get safety right is about compassion, not operating profits or personal success. This fact has very interesting implications for working on safety improvement with leaders at all levels.

Although "cultural unity" is usually a secondary motive, it can be critically important. Achieving real unity in the organizational culture is very difficult for most companies. Even leaders in this area recognize that they have a long way to go. Organizational life is inherently challenging. Often the organization's most important goals seem disconnected from the worker and the supervisor, if not the manager. Organizations sometimes impose a set of "values" on their employees to try to improve performance. Employees often view such efforts with cynicism. *Highly effective leaders recognize that taking a leadership role in safety gives them an opportunity to create real shared values in the organization.* When done carefully, this has remarkable effects on organizational citizenship, the ability of employees to work effectively as teams, and to overall organizational effectiveness. We will discuss this in depth in Chapter 4.

Nothing undermines the effort to create cultural unity faster than a workplace that is perceived to be unsafe.

In some companies, highly significant costs are associated with injuries. However, in our experience, injury costs are more relevant to the justification of needed resources than to motivating leadership. Further, improving safety to reduce costs can be taken wrongly by front-line employees if they think it is the only reason leaders want to make safety improvements.

Influencing the Behavior of Safety Leaders

Organizational leaders vary in their abilities and skills to provide safety leadership. This is as true for the senior leader as the first-line supervisor. Some leaders have a natural inclination towards safety and need little help; others are quite reluctant to take on safety issues, and may even be apprehensive about it. Leaders are often chosen for their technical ability, and providing excellent safety leadership is necessarily a "people activity" requiring high levels of interpersonal skills.

We will look at these issues in depth in the next three chapters as we examine the Safety Leadership Model. A summary of the main points that address this issue follows:

1. Safety leadership behaviors are subject to the same principles as any other set of behaviors. *However, doing the right things to influence behavior may vary widely from the senior leader to the manager, supervisor, or worker.*

2. Great safety leaders are great leaders who are motivated to improve safety; they are no different than great leaders generally.

3. We know the personality characteristics, leadership styles, and best practices of great safety leaders. We can describe the specific behaviors and practices that are necessary to provide safety leadership and culture change. And we can specify the attributes of a healthy organizational culture and safety climate and how to measure them.

Sustaining Organizational Change: Two Critical Elements

Every leader has seen initiatives that fail to sustain a desired change. In fact, it is more usual than unusual for change initiatives to run their course in a few years, leaving only a small effect on the performance area intended for long-term change.

We think this unfortunate fact of organizational life is unnecessary if the change effort is approached properly. There are two elements that distinguish short- and long-term change efforts in safety performance:

1) Mechanisms and Processes

2) Serious Employee Engagement

It's difficult to overstate the importance of these two elements. Most leaders have grown accustomed to mediocre change efforts and have come to expect them to have minimal impact. But this need not be the case if the strategy addresses these two critical elements adequately. "Mechanisms" and "engagement" are active words, not descriptions in manuals that describe how things are supposed to be. They refer to sets of behaviors that are performed routinely as part of day-to-day operations that make the crucial difference between "training programs" and real organizational change.

Mechanisms and Processes

A mechanism is a set of ongoing activities undertaken to create an organizational change. Say we want to improve the way supervisors relate to the Working Interface. One strategy is to train them and then hope that what they learn will be applied in their day-to-day work. This isn't likely to work well because it leaves them on their own to figure out the most difficult aspect of the change — how to integrate what they learned into what they do every day.

Another strategy is to create a system that requires supervisors to perform regular actions, and uses a data tracking system to follow its results and provides organizational leaders with measured data points. For example, supervisors could be required to look systematically at the Working Interface, make notes and enter them into a data system, and produce summarized monthly reports for senior leaders. Another system, described in Chapter 9, would have workers participate in keeping the Working Interface free from hazards. And the behavior observation and feedback system NASA adopted to improve the behavior of its leaders, described in Chapter 12, would be still another.

Mechanisms require that people get involved to operate them. This leads to the second element: serious employee engagement.

Serious Employee Engagement

The closest thing to magic in organizational change is getting the employees excited about what is going on. And the most effective way to do this is to involve them, to give them actual responsibilities in making the mechanisms and processes work. It's a lesson organizations have to learn over and over, and they still tend to forget it. There is no substitute for employee engagement.

Most organizations learned this lesson in the '80s and '90s doing quality improvement. But as other changes became necessary — new leaders, new technology, new challenges — the lesson was lost to many. In our experience with NASA's culture change effort, this lesson was brought home. Each aspect of the intervention plan was effective, but what really caught the attention of both leaders and individual contributors, what told everyone in the organization that things were really changing, was the fact that employees got involved. Ironically, this involvement could well have been lost because it required so much time from employees — time no one had. But, like physical exercise, doing more ends up being less. As the body gets in shape, new energy emerges.

Throughout this book, we will maintain this central orientation: organizational change requires that the concepts and theories developed to support the change be put into specific ongoing mechanisms and processes. Employee engagement is both a way to keep processes active as well as being a process itself.

• • • • •

Section 2

The Safety Leadership Model

In Chapter 1 we looked at the core elements of organizational safety performance. While each element is important, organizations usually find the leadership element most challenging. When safety is given new emphasis, organizational leaders often wonder what they can and should do differently — "We want to do the right things to improve safety performance, but we aren't really sure what they are." The three chapters in this section address the question "What does it take to be a great safety leader?"

Chapter 2

The Safety Leadership Model, Part 1
The personality, values, emotional commitment, and leadership style of the effective safety leader

This chapter was written in collaboration
with John Hidley, M.D.

- The core elements: Personality, values, and emotional commitment

- Measurement of the Big Five

- Applications of Big Five research to safety leadership

- Using the findings to improve safety leadership

- How leaders use the Big Five to improve safety effectiveness

- The leader's values and emotional commitment to safety

- Leadership style: transactional and transformational

- Cultivating style

Figure 2-1. The Safety Leadership Model.

Introduction

When an organization focuses on safety improvement, the role of the individual leader becomes increasingly apparent. Organizational leaders who wish to be great safety leaders ask, *"How do we lead safety? What exactly do great safety leaders do?"*

In this chapter, we will begin to look at a comprehensive model that provides a clear understanding of the critical elements of safety leadership and how these elements relate to each other. Together with the Organizational Safety Model shown in Chapter 1, this leadership model provides the leadership team with a broad, empirically-based understanding of what it takes to provide successful safety leadership, and what it looks like when achieved.

The model is multi-dimensional (Figure 2-1) and can be thought about in two ways: either as emanating outward from the center, from the individual's core values to the group; or from the outside in, from the culture of the

26

organization to the core values of the individual leader. Each of the rings of the model represents a leadership dimension that contributes to successful safety leadership. The outer ring considers the results of effective leadership on the culture of the organization by presenting the specific elements of organizational culture that are critically important to safety excellence.

The model forms the basis for assessing both the leader's and the organization's current state with respect to safety leadership and organizational culture. It can be used with the individual leader, the leadership group, or the organization as a whole.

As we built the model, we required that any proposed dimension be measurable and empirically validated before it was included. Therefore, the model as a whole provides the leadership team with a profile of where it currently is, and where it needs to go to achieve its objectives. Because extensive data has been gathered on each ring, an individual leader or organization can use it as an assessment tool, contrasting themselves with other safety leaders and organizations worldwide.

Core Elements –
Personality, Values, and Emotional Commitment

In this chapter we will discuss the inner two rings of the model — the core elements of safety leadership in the individual. This includes the personality, values, and emotional commitment of the leader, as well as his or her leadership style. In Chapter 3 we will review leadership best practices and in Chapter 4 we will look at organizational culture.

The Leader's Personality

Personality refers to individual differences in how a person tends to think, feel, and act. We are all different, but what tendencies or dispositions make us different? Personality research addresses what is unique about the individual in terms of specific traits and attributes.

Psychological research in personality has been going on for at least the last fifty years. Hundreds of studies have been published which seek to define

FACTOR	CORE CONCEPT	SAMPLE DESCRIPTORS	RELEVANCE TO SAFETY LEADERSHIP
Emotional Resilience	How stable are they emotionally?	How free are they from stress, anxiety, worry, frustration, irritability, anger, moodiness and feeling overwhelmed? How easily are their feelings hurt?	A degree of stability is required for successful relationships and perspective, but in the extreme can result in complacency.
Extroversion	How outgoing, talkative, positive and dominant are they?	How assertive are they? How easily do they mingle and make friends? How comfortable are they taking charge and being the center of attention?	Extroversion provides the basis for the alliances needed to foster good relationships but in the extreme can produce expansive-but-empty visions and callous, self-defeating behaviors.
Learning Orientation	How extensive is their imagination and mental life?	How effectively and rapidly do they generate new and interesting ideas, new approaches? How much information can they handle and how quickly do they understand things? How rich and difficult is their vocabulary and do they take the conversation to a higher level? Do they seek challenging information?	This trait can assure that the creativity and new ideas needed for change will be there, but if it is unmoderated, it can put the leader too far out ahead of those he must influence.

Table 2-1. The Big Five as relevant to safety leadership.

28

FACTOR	CORE CONCEPT	SAMPLE DESCRIPTORS	RELEVANCE TO SAFETY LEADERSHIP
Collegiality	How interested in, and sensitive to, the needs and feelings of others are they?	Do they have the capacity for gratitude, empathy and consideration of others? Do they have the ability to relate to others and make them feel at ease? Do they have the ability to get along with, support, coach and help others?	This assures they possess the sensitivity to other individuals that is needed for compassion and to help them grow. At the extreme, it may make the leader insufficiently demanding.
Conscientiousness	How important to them is a structured and reliable approach?	Do they need regularity, a schedule and punctuality? Do they attend to details? Are they well prepared? Do they have a plan and the ability to stick to it?	This trait leads to effective attention to detail and an orderly process of task implementation. At the extreme, it can lead to too much task focus at the expense of relationships.

Table 2-1. The Big Five as relevant to safety leadership (continued).

measurable personality traits. A wide variety of personality inventories and typologies have been developed, with varying degrees of success. While somewhat useful in certain situations, a comprehensive grasp of personality traits and attributes eluded researchers for many years.

This research took a giant step forward when computer-based factor analysis (a sophisticated statistical technique) revealed a very interesting finding. Of the dozens of personality attributes and traits, factor analysis showed that they could be reduced to five core attributes that define individual differences.

Known as the "Big Five," they have been subjected to extensive research over the last ten years or so.

This cluster of five underlying factors has been shown to hold across people generally, even across cultures, and to persist across time. This research represents a major step forward for the psychology of personality. Today the Big Five factors are regarded widely as the underlying dimensions of personality. Research on the Big Five has yielded information that has proven to be very predictive, especially in the assessment of effective leadership.

Table 2-1 shows the Big Five factors and briefly points out some of the ways they are relevant to the personalities of safety leaders.

Measurement of the Big Five

When you notice the ways in which the people you work with are different, the things you are noticing are very likely to be Big Five dimensions. What is unique about the personality of the individual is primarily captured in the Big Five.

Personality assessment instruments have been developed which measure these five essential aspects of personality. Extensive research has found them valid and predictive of leadership success. Our task in improving safety leadership is to adapt this research to assist safety leaders to become more effective.

Why is it important for leaders to know where they stand on each of these dimensions? The scientific literature shows that scores on these factors correlate with various aspects of leadership and career success.[1,2,3] Although individual leaders are unlikely to change their personality, they can certainly adopt critical behaviors that compensate for personality attributes, and improve their effectiveness.

---

[1] Hogan, R., G.J. Curphy, and J. Hogan. 1994. "What We Know About Leadership: Effectiveness of Personality." *American Psychologist*. 49: 493-879.

[2] Hurtz, G.M. and J.J. Donovan. 2002. "Personality and Job Performance: The Big Five Revisited." *Journal of Applied Psychology*. 85: 869-879.

[3] Mount, M.K., M.R. Barrick, and G.L. Stewart. 1998. "Five-Factor Model of Personality and Performance in Jobs Involving Interpersonal Interaction." *Human Performance*. 11: 145-165.

Applications of Big Five Research
to Safety Leadership

A substantial body of research has been published that provides data relevant to safety leadership improvement. Big Five variables have been shown to predict both leadership effectiveness and leadership emergence (who will emerge from the group as the leader).[4]

For leadership effectiveness, Emotional Resilience, Extroversion, and Learning Orientation showed statistically significant correlations that generalized across studies. In addition, research showed that Collegiality was related to leadership effectiveness, although the correlations were not as strong as for other attributes. As for leadership emergence, Extroversion and Conscientiousness were found to be the strongest predictors. Not surprisingly, Collegiality was found to be predictive of success in jobs that require significant interpersonal interactions.

As an illustration of how robust the Big Five are, one study gathered data on 244 families over fifty years and found that Big Five measures taken in childhood had statistically significant predictive capability of adult career success.[5] This may not be surprising to those who have raised children to adulthood and recognized certain attributes in young children that persisted into adulthood and contributed to their career success. But it is another thing to demonstrate these relationships scientifically.

What bearing does all this have on safety leadership? There is a direct as well as an indirect relationship. Leadership effectiveness and leadership emergence are both highly relevant to the selection and development of safety leaders. Effective leaders generally will tend to make effective safety leaders.

It isn't difficult to see how each of the Big Five attributes has a direct relationship to safety leadership effectiveness. (Table 2-2)

[4] Judge, T.A., J.E. Bono, R. Ilies, and M.W. Gerhardt. 2002. "Personality and Leadership: A Qualitative and Quantitative Review." *Journal of Applied Psychology.* 87: 765-780.

[5] Judge, T.A., C.A. Higgens, C.J. Thoreson, and M.R. Barrick. 1999. "The Big Five Personality Traits, General Mental Ability, and Career Success across the Life Span." *Personnel Psychology.* 52: 621-652.

FACTOR	LEADERSHIP IMPACTS
Emotional Resilience	High resilience may give a leader a cool head in the midst of a safety crisis. On the other hand, the leader may appear aloof to others. With very high resilience, the leader may be impervious to feedback. Low resilience may result in strained interpersonal relationships and difficulty enlisting others in safety goals and objectives.
Extroversion	An extroverted safety leader is more likely to be out on the floor or otherwise in contact with people about safety. Engaging others in safety interactions is critical for safety leadership, and the extroverted leader finds this natural to do. In the extreme however, extroversion can be seen as superficiality. An introverted safety leader may not be easily accessible to others, and will have to find ways to compensate for this tendency.
Learning Orientation	The leader with an academic learning style will find books, articles and lectures on safety interesting and stimulating. He or she will tend to approach others on the same basis, but this may or may not be their preferred style. Similarly, the leader whose style favors learning on the job will face analogous challenges when trying to train or inspire those with a more academic style.
Collegiality	A highly collegial safety leader will tend to have the compassion needed for safety motivation. At the extreme, however, such a leader may not be sufficiently demanding. A leader with low collegiality will need to find ways to compensate and devise other kinds of personal motivation.
Conscien-tiousness	A highly conscientious safety leader will be naturally inclined to attend to the many details necessary to high-level safety performance. At the extreme, he may neglect the big picture and become too involved in the day-to-day concerns and allocate all his attention to compliance issues. A safety leader with low conscientiousness may have grand ideas, but little credibility.

Table 2-2. Big Five impact on safety leadership effectiveness.

Using the Findings to Improve Safety Leadership

These findings all suggest that the personality structure of the leader is important to his success. But if the personality you had as a child impacts your leadership fifty years later, does that mean you are limited or doomed by the personality you were born with?

While it is true that you probably cannot change your personality, even if you wanted to, fortunately it is not even necessary. There are great safety leaders among a variety of score patterns on Big Five attributes.

How can that be if personality attributes are predictive of leadership performance? The data show, for example, that Extroversion has the strongest relationship to leadership emergence. If you imagine a great safety leader, likely you are not picturing someone who sits in his office all day and only interacts with his computer. Nevertheless, there are great safety leaders who do not score especially high on Extroversion. They are able to perform so successfully as safety leaders because they have learned to compensate behaviorally for their low Extroversion. They can't change their personality but they can change how they behave and interact.

Low Emotional Resilience correlates with leadership difficulties and low career success because it can impact thought processes and therefore decision-making. It may also result in relationship difficulties. The important point, however, is that it does not have to result in these problems. A person with low Emotional Resilience can learn to compensate with self-management behaviors that others may not need to learn.

This tells us that we all have attributes that make certain skills and capabilities come to us naturally. But the absence of such attributes doesn't have to limit us; it just requires our willingness to learn new behaviors. Knowing your own scores on the Big Five will give you insight into the kind of leader you tend to be, and where you need to compensate behaviorally to be more effective.

Learning your scores also provides guidance as to what you should and should not try to change. For example, if you score low on Extroversion, you would be fighting an uphill battle against your own nature to try to turn yourself into an outgoing, gregarious people-person. Being the life of the party is not who you are. On the other hand, it might be relatively easy and critically important for you to spend a little additional time talking one-on-one with each of your reports. Leaders low on Extroversion exert leadership influence through their relationships, just as extroverts do, but they may be more successful doing it one-on-one.

How Leaders Use the Big Five to Improve Safety Effectiveness

A senior VP at a large western oil company was faced with the necessity of making a major reduction in his workforce. He wanted the change to be received as positively as possible and he did not want a negative impact on safety. He had planned to hold special all-employee meetings at each of the four refineries in his division to provide important information about a long anticipated (and dreaded) reorganization, which would result in a reduction in the number of employees. He also planned to use these meetings to speak about the need for increased safety vigilance during the reduction.

This leader was highly intelligent and motivated on behalf of safety excellence. He was technically brilliant and well respected at all levels within the organization. He knew that getting the right message across to personnel throughout the refinery was critical to the success he needed. He had seen other leaders speak to large groups of employees and do the same thing he needed to do. They would speak from the heart and give a moving yet crisp account of the situation and why the changes were necessary. They would then win the groups' assent, if not their enthusiasm. He knew what needed to be done and he was motivated to do it well.

Nevertheless, he dreaded the task and did not feel a high level of confidence that he could do it as well as necessary. He had always been uncomfortable with all-employee meetings and had grown in his responsibilities despite this aversion because of his other strengths. What do you think his Extroversion score was? Of course, it was very low. Speaking effectively to large groups had never come to him naturally, and this situation called for him to speak about a difficult and unpopular issue.

If you had been this leader's coach, what would you have suggested? Should he have forced himself to try to be a charismatic speaker? Should he have asked someone else to do the talk?

The VP talked over the situation with his coach and came up with a small but pivotally important refinement to his communications plan. He decided that the all-employee meetings would be brief and cover only the aspects of the ensuing changes that everyone needed to hear at the same time. He would also announce that over the next several days he and the refinery manager would spend all their time talking with individuals and small groups in the refinery about details of the changes and his concern for people's safety.

The leader was much more comfortable with this approach. He was more able to "be himself" one-on-one and in small groups and therefore get his messages across more convincingly. As it turned out, he was very successful not only at helping to facilitate the changes, but also in assuring a safe transition.

Knowing what you should compensate for behaviorally and understanding what you should not try to change are both significant benefits of having an individual assessment based on the Big Five. This kind of self-knowledge also fosters a leader's capacity for relationships. This is particularly important for safety leadership, which requires compassion, something not as important to other types of leaders. And compassion grows with the capacity for real and meaningful relationships.

A leader's insight into himself as a human being with unique strengths and foibles gives him an opportunity to cultivate empathy and compassion. These are necessary qualities for a great safety leader. And cultivating them does not mean giving anything up. On the contrary, it means gaining maturity and a more profound sense of humanity.

These considerations bring us to the next aspect of the inner ring of the model — values.

The Leader's Values and Emotional Commitment to Safety

The word "value" expresses the notion of worth or desirability. There are two categories of value: intrinsic and extrinsic. Intrinsic values have worth for their own sake; they are ends in themselves and have ethical import because they characterize what we think people *should* be pursuing. Extrinsic values, on the other hand, have worth only as a means to an end and their import is in their utility.

Getting promoted, having personal power, and acquiring prestige are good examples of extrinsic values. Job titles, power, and prestige are all rightly regarded as having worth because they are useful. Human life, ethics, a sense of duty or stewardship, and the well-being of the individual are examples of intrinsic values. They are valuable in themselves, not because of their utility. One can't say that a person has an ethical obligation to become powerful, but one can say that people *should* value human life and well-being.

The worth of extrinsic values is derivative. They get their worth because of their power to further intrinsic values. The reason money is valuable is because it can further one's happiness and well-being, for example.

A good leader is sensitive to extrinsic values. This is how he or she keeps the organization focused and working to achieve its proper end — attaining objectives and maximizing the bottom line. In addition, however, a great safety leader is also sensitive to intrinsic values. He believes in and is deeply committed to the worth of the individual. And this belief is deeply felt as an emotional commitment to the health and safety of each individual employee.

One senior leader of a 25,000-employee manufacturing organization described those feelings this way: "When I was a Vice President of Operations, my children were college-aged and living away from home. When the phone would ring unexpectedly, sometimes in the early morning hours, the first thing I thought of was them. Were they okay, or was there a problem prompting the unexpected phone call? Fortunately, I've never had a call that told me my children had been harmed. But I have had phone calls saying that our employees have been harmed. And I think about it the same way each time I get one of those phone calls. We are a family here in this organization, and my commitment to our employees is like my commitment to my family members."

A great safety leader can have this kind of emotional commitment because he doesn't relate to the organization's employees as merely a resource or as an anonymous group. If the company is large, a senior leader can't know all employees individually, but he is acutely aware that every one of them is an individual human being like himself who experiences life as intrinsically valuable. And he has sufficient empathy to respect that fact.

Being an effective safety leader takes something over and above what it takes to be a good leader generally, and it is this awareness and feeling, this emotional commitment that makes the difference. It requires a significant degree of empathy, compassion, and maturity. These qualities are available to all leaders, but they must be cultivated and nurtured. Most people have such qualities, but many don't know how to allow them to interact effectively with their business leadership roles and personas — allowing these qualities to be influential to the right degree and visible to the right extent.

Why is this important? Why is it worth the effort? Because what we value is what we strive to achieve, and a leader's values color and shape the direction set for the organization. The leader's values play out in the culture he or she creates. For example, if a leader relentlessly and exclusively pursues extrinsic values (or even if that is just how employees perceive it), an organizational culture may be inadvertently created in which employees cut corners in the name of production or profitability. At the very least, a culture is created in which employees are not taken care of properly and the organization's safety performance plateaus, allowing serious injuries and fatalities to continue.

Intrinsic values serve as a kind of behavioral and cultural insurance policy both for the organization and the individual employee. They are a boundary that keeps both out of trouble.

But there's more to it. Because extrinsic values derive their worth from their relation to intrinsic values, a leader's neglect of those values ultimately disconnects employee behavior from its motivational source. People don't want to work *just* for the money. They want to work at something they value doing. If a leader neglects intrinsic values, or worse, if he creates a conflict between extrinsic and intrinsic values, employee morale and motivation suffer. Conversely, if a leader calls upon employees' intrinsic values, greater engagement will result.

So, it is important for a leader to understand what he actually values. Compassion for others is a basic core emotion found in almost all of us. But the pressures and frustrations of day-to-day organizational life may drown it out. For some leaders this critically important compassion comes naturally, for others it doesn't. In the latter case, it's the job of the senior-most leader in the organization to awaken it in others.

Knowing what we really value, and really valuing safety, is necessary but not sufficient to create effective safety leadership. It is also true of human nature that we often do not see ourselves as others see us. You may feel that you are a compassionate individual who values your fellow employees. And you may really be that kind of person. But others don't judge you according to how you feel about yourself, or even by your intentions. You are judged by your behaviors, the visible things you do and say, the decisions you make, and the way you communicate, or fail to communicate about them.

Chapter 3 explores the specific behaviors that great safety leaders engage in. But first, we'll examine the importance of leadership style.

Leadership Style

Over the years, leadership style has been classified in a number of ways in the research literature. In recent years, various dimensions have coalesced into two basic styles: *Transformational Leadership* and *Transactional Leadership*. (A third type, *Laissez-Faire Leadership,* is also referred to, but this really amounts to an abdication of leadership responsibility.)[6] There is increasing evidence that transformational and transactional leadership are not mutually exclusive, but that different situations call for different styles and that great leaders are adept at using the mix that is appropriate to a given situation.[7]

Transactional Leadership

This style concerns the connection between performance and rewards, and posits that both employees and leaders are motivated by self-interest. The word "transactional" refers to the *quid-pro-quo* nature of the relationship between the leader and his or her followers. A good transactional leader creates conditions that coordinate his self-interest with that of the employees.

Transactional leadership can be active or passive. In the former the leader actively communicates expectations and then monitors and reinforces performance. The research literature calls this *constructive* transactional leadership. In the passive version, the leader waits until something goes wrong and then

[6] Antonakis, J., A. T. Cianciolo, and R. J. Sternberg, (Editors). 2004. *The Nature of Leadership.* Thousand Oaks: Sage Publications.

[7] Avolio, B. J. 1999. *Full Leadership Development: Building the Vital Forces in Organizations.* Thousand Oaks: Sage Publications.

responds with the appropriate consequence. This is called *corrective* transactional leadership or *management by exception*. The literature very strongly demonstrates the superiority of constructive over corrective transactional leadership, but says that few leaders avail themselves of its power.[8]

All transactional leadership, which has also been called *task-oriented* leadership, is essentially conservative. It is an important leadership style for preserving existing cultural conditions and organizational practices and processes. It aims to get things done within the current context and works best in stable environments.

Transformational Leadership

This leadership style focuses on the future and is essentially developmental. It is most valuable when the task involves creating order out of chaos, breaking deadlocks, creating significant change in the organization, or developing future leaders — goals to transform the organization and its employees. It has also been called *relationship-oriented, charismatic,* or *inspirational* leadership. A transformation leader's role is to inspire employees to go above and beyond their mere self-interest.

The Conference Board's *CEO Challenge 2004* lists the challenges that 539 CEOs from around the world named as their top ten concerns (of the sixty-two from which they could choose). Five of the top six are clearly situations that call for strong transformational leadership skills, and success in virtually all these challenges is related to that style of leadership.

1. Sustained and steady top-line growth

2. Speed, flexibility, adaptability to change

3. Customer loyalty/retention

4. Simulating innovation/creativity/enabling entrepreneurship

5. Cost/ability to innovate

---

[8] Avolio, B. J. *Op cit.*

6. Availability of talented managers/executives

7. Tight cost control

8. Succession planning

9. Seizing opportunities for expansion/growth in Asia

10. Transferring knowledge/ideas/practices within the company

Clearly, this kind of leadership is critically important in today's environment and it is especially important for safety. Several studies show that high transformational leadership is a good predictor of good safety performance.[9] This performance enhancement is mediated by the aggressive, concrete actions that transformational leaders take to address identified safety concerns and issues.

Because of the differences between the two styles of leadership, the virtues and behaviors that characterize them are different. Key differences are shown in Table 2-3.

Which style is more desirable for safety leadership? In the research literature, transformational leadership has been shown to predict a substantially higher level of performance than transactional leadership.[10] As you might guess by examining the differences between the two styles in Table 2-3, it does this because it mobilizes more employee energy and enthusiasm by linking an employee's intrinsic values with the organization's vision. In transformational leadership, employees are motivated by their personal, intrinsic values, not just by a paycheck. The key to this type of leadership is the effective appeal to the employees' intrinsic values.

It turns out, however, that you cannot be an effective transformational leader without strong transactional skills and that both styles are necessary for great safety leadership.

---

[9] Barling, J., C. Loughlin and E.K. Kelloway. 2002. "Development and Test of a Model Linking Safety-Specific Transformational Leadership and Occupational Safety." *Journal of Applied Psychology.* 87: 488-496.

[10] Avolio, B. J. *Op cit.*

	TRANSACTIONAL LEADERSHIP	TRANSFORMATIONAL LEADERSHIP
Ethics	Is highly individualistic with an ethic of self-interest; seeks mutual advantage through contractual relationships and fair play.	Is highly social with an ethic of the greatest good for the greatest number; seeks to resolve value conflicts through win-win strategies and balancing the needs of all constituencies.
Motivation	Emphasizes the connection between performance and rewards; e.g., creates explicit expectations, monitors and provides feedback.	Creates enthusiasm for the vision and loyalty to the leader, the organization, and the envisioned future; e.g., creates and passionately communicates a meaningful vision. Models and inspires.
Scope of Work	Negotiates the performance/ reward equation.	Expects people to go above and beyond their self-interest for the good of the group and exemplifies this in personal behavior.
Relationship	Task-focused, is reliable and fair but not necessarily personal.	Person-focused, is personally involved with employees based on their individual needs; e.g., makes an effort to help them achieve their aspirations. Over time, turns followers into leaders.
Emphasis	Emphasizes getting the job done; does not encourage initiative or going outside the box; sees failures as impediments to production.	Encourages initiative and actively challenges old ways; emphasizes finding new and better ways to do things; sees failures as learning opportunities.
Communication	Does not provide more information than employees need to know to do what is expected of them.	Shares big-picture information widely and encourages others to communicate and express their opinions.
Employee Impact	Big difference between the corrective and constructive methods with the latter preferred by employees.	Generally followers prefer transformational style; lower turnover.

Table 2-3. Differences between leadership styles.

Transactional Safety Leader Activities

The leader needs to make expectations and priorities very clear, actively monitor compliance, and reinforce successes. For example, many elements of the safety system need to be preserved and strengthened. The leader needs to ensure that all of the organization's enabling and supporting safety systems are functioning well, and he may need to do it personally by reviewing audit data about these systems and addressing the results systematically and effectively. If the leader is the CEO, he may need to ensure that his reports are on top of such issues. These activities all call for a transactional style and transactional behaviors.

Transformational Safety Leader Activities

Safety leaders are also called upon to build a strong safety culture, and this inevitably involves cultural change and organizational development. It may also require developing increased bench strength in safety leadership at various levels of the organization. The leader may need to create a vision of the strategic role that safety plays in the organization's future, challenge complacency, and develop leaders who can implement the cultural changes needed to realize the leader's vision. These activities are clearly within the domain of the very senior-most leaders and demand a transformational style of leadership.

Research also shows that transformational leadership is more effective than a transactional approach when the leader has little or no control over how the employee will be rewarded for a satisfactory performance. This finding is critical to safety leadership, where this is very often the case. Workers usually work without continuous direct supervision and are often subject to many pressures that predispose them to take shortcuts. If they do something in an unsafe way, the leader may only see that they got the job done, not how safely they did it. The higher the leader, the more acute this problem becomes.

Cultivating Style

Unlike personality, style is something a leader can learn. Style is a matter of *how* he approaches opportunities, what he generally focuses on, what he ignores, what he chooses to emphasize, and what he delegates to others. These things are all behavioral and within the control of the leader.

If a leader wants to improve his capacity for either of these two leadership styles, it is important that he clearly understands his natural inclinations. Personality factors predispose to stylistic preferences but they do not determine them. For example, Extroversion impacts safety performance by predisposing a safety leader to a transformational style. Leaders who do not score high on Extroversion can nevertheless learn to be very strong transformational leaders. They may have to make more of an effort and they will build on other facets of their personality but they can be just as effective.

Transformational Leadership Style is usually measured by a 360-degree diagnostic instrument. Four variables are usually measured:

- **Challenging** – Provides subordinates with a flow of challenging new ideas aimed at rethinking old ways of doing things; challenges dysfunctional paradigms; promotes rationality and careful problem solving.

- **Engaging** – Helps others to commit to the desired direction; coaches, mentors, provides feedback and personal attention as needed, and links the individual's needs to the organization's mission.

- **Inspiring** – Sets high standards and communicates about objectives enthusiastically; articulates a compelling vision and communicates confidence about achieving the vision.

- **Influencing** – Builds a sense of mission-beyond-self-interest and a commitment to the vision; gains confidence, respect and trust; considers the ethical consequences of decisions; appeals to other's most important values and beliefs; instills pride; models these kinds of behaviors.

Using a 360 diagnostic instrument will give the leader a picture of where to focus to strengthen his or her transformational leadership. Actually doing so requires focusing on the appropriate leadership best practices, which are discussed in the next chapter.

• • • • •

Chapter 3

The Safety Leadership Model, Part 2
Best practices in safety leadership

- The central role of leadership in safety

- Leadership vs. management

- Best practices in safety leadership:
    Vision
    Credibility
    Action orientation
    Collaboration
    Communication
    Recognition and feedback
    Accountability

- Measuring safety leadership best practices (and leadership style)

The Central Role of Leadership in Safety

In the preface to this book, we referred to our research which found that leadership was a central variable predicting the success of safety initiatives. In the Introduction and Chapter 1 we discussed how safety can be a metaphor for organizational excellence, how it is possible to Lead with Safety, and we gave examples of well-known organizational leaders who have done so. In Chapter 2 we dealt with the personality, values, and emotional commitment of the leader, as well as his or her leadership style.

Now we turn to the specific best practices that comprise safety leadership itself: *What do great safety leaders do? Is being a great safety leader different than being a great leader generally, or is it the same?*

We have all known great safety leaders, people whose commitment to safety, combined with excellence in leadership, has enormous positive influence in making the organization safe. This chapter is about what those people do that distinguishes them as great safety leaders. Keep in mind that these people are needed at all levels of the organization. They can be formal or informal leaders. If we understand what they do in concrete behavioral terms, we will be able to develop other leaders like them throughout the organization.

All this is based on the premise that *improving* safety in the organization is essentially about leadership creating a strong organizational culture and safety climate. Within this environment, enabling safety systems thrive, sustaining safety systems are held in place, and the Working Interface is continuously made safer through the reduction of exposure to hazards.

The most difficult aspect of safety improvement is not the implementation of safety systems and mechanisms. These things are essential, but they are relatively easy and can be managed. The difficult part is creating a culture in which safety is a driving value. And this is where leadership comes in. Creating that kind of culture, or its opposite, is something done by leaders.

Figure 3-1. Leadership vs. management.

Leadership vs. Management

Many distinctions have been made between leadership and management elsewhere so we will only give a brief summary here. For this book the distinction intended is between the task-oriented behaviors of getting things done, and the way in which the leader performs those tasks.[1] Figure 3-1 illustrates this distinction. Managing behavior tells other people what to do: schedule training events, perform jobs at particular times, start work now, and stop later. It engages the minds of workers and causes them to take action.

[1] Kouzes, J. and B. Posner. 1995. *The Leadership Challenge.* San Francisco: Jossey-Bass, Inc.

Leadership, on the other hand, has more to do with *how* the task to be performed fits the overall goals of the organization. It engages the hearts of workers and affects their level of motivation, their connection to the work.

Organizations that function best in safety have effective leaders as well as managers — people who not only direct what is to happen next, but who also get subordinates to understand why things are done the way they are, why it matters to the larger goals of the organization, and why it is important.

Best Practices in Safety Leadership

In working with safety leaders across many organizations, we have identified a set of best practices, behaviors that effective safety leaders engage in that create a strong organizational culture and safety climate.

Seven Best Practices
for Excellence in Safety Leadership

1. Vision
2. Credibility
3. Action orientation
4. Collaboration
5. Communication
6. Recognition and feedback
7. Accountability

These practices are valuable to leadership generally, but we have derived them with safety in mind. Effective general leaders are effective safety leaders if they value safety adequately. Effective safety leaders tend to be effective leaders in general. Developing leaders who learn to lead safety well will benefit the organization in many ways.

It is useful to think about these practices as sequentially related; one builds upon the other. The leader who wants to develop best practices would be wise to consider them in this way. We will first describe each practice and then discuss how to measure it.

Safety leadership starts with *vision* on the part of the leader:

I can see the future state and I can describe it in a way that others find compelling. My effectiveness depends in large part on the degree to which others perceive that I have credibility. No one will take my vision seriously unless they find me credible. Only if I have a record of consistency, am known for keeping my word, and stick with the truth even when it isn't popular, will the things I say about safety in the future have real meaning and be influential.

If I have a vision and can talk about it effectively, and if I am credible, I also need to have a strong action orientation. When I see the facts clearly, I make the decisions that follow from them. I do not hesitate to take decisive action. I actively collaborate with others in my organization around safety issues. I consult with others before making final decisions that affect them. I realize that to be effective I have to foster excellence in communication about safety issues.

Having set the stage for safety excellence, I stand ready to provide feedback and recognition when I see changes in the desired direction. I notice when people change their behavior and I let them know that I notice and appreciate the change. Finally, I know the value of accountability, and I hold people accountable for the results that must follow when the right things are done.

Note that the first two practices — vision and credibility — are the foundation upon which the others are built, and both are primarily *antecedent* in behavior analysis terms they are triggers of behavior (see Chapter 5). Action orientation, collaboration, and communication are also primarily antecedent, and are process related, ongoing maintenance of the many things the organization has to accomplish. The last two practices concern *providing consequences*. These practices follow the behaviors that are required for improvement, and they reinforce them.

Vision

· Acts in a way that communicates high personal standards in safety

· Helps others question and rethink their assumptions about safety

· Communicates the organizational vision through word and action

· Demonstrates willingness to consider and accept new ideas

· Helps people consider the impact of their actions on the safety of others, and on the organization's safety culture

· Challenges and inspires people around the safety vision and values

· Describes a compelling picture of what the future could be

Figure 3-2. Behaviors that demonstrate Vision in safety.

1. Vision

The great safety leader has a clear picture of the future state of safety and articulates that picture in a compelling way. Vision is a quality many leaders lack or find difficult. Others seem to have it naturally and take it for granted. The essence of a vision for safety is being able to "see" what the future desirable state looks like. How is it different from the way things are today? What kinds of things will people do and say that they don't today? What decisions will be made differently, and what assumptions underlie those decisions? If by some miracle you were able to change the organizational safety culture and climate today, what would you see tomorrow that would be different? The effective safety leader needs to be able to see these things vividly and to describe them in compelling terms.

What makes the description of the vision compelling? In part, it's the ability of the leader to describe it plausibly, with enthusiasm and excitement. In part, it's the personal credibility of the leader.

Credibility

- Admits mistakes to self and others

- "Goes to bat" for direct reports; represents and supports the interests of the group with higher management

- Gives honest information about safety performance, even if it is not well received

- Asks for ideas on how to improve his or her own performance

- Acts consistently in any setting and applying safety standards

- Is willing to make safety-related decisions that are unpopular or involve some personal risks

- Demonstrates personal concern for employee well-being

- Follows through on commitments made

- Treats others with dignity and respect

Figure 3-3. Behaviors that demonstrate Credibility in safety.

2. Credibility

Great safety leaders have high levels of credibility, with direct reports and with the larger organization. People believe what they say and trust them to tell the truth, even if it is unpopular and unlikely to be well received. They are known to be free from personal agendas. Most importantly, their actions are seen by others to be consistent with their words — being consistent is not enough; effective safety leaders must also be *perceived* as consistent.

Action Orientation

· Is proactive rather than reactive in addressing safety issues

· Gives a timely, considered response to safety concerns

· Demonstrates a sense of personal urgency and energy to achieve safety results

· Performance driven — delivers results with speed and excellence

· Focuses safety efforts on the most important priorities

· Shows persistence in solving safety problems

· Does whatever it takes to make safety improvement initiatives successful

· Seizes safety improvement opportunities when they arise

· Is creative and innovative in improving safety

Figure 3-4. Behaviors that demonstrate Action Orientation in safety.

3. Action Orientation

The great safety leader needs to be perceived as willing to take action on behalf of safety issues when it is appropriate. This reinforces credibility and tends to flow naturally from it. Safety issues arise and decisions must be made: shut down the process or continue with it, do the maintenance task now or later, spend the resources necessary to address the hazard at its source or do something temporary to mitigate the problem. Action orientation means persistence, innovation, and personal urgency.

Collaboration

- Promotes cooperation and collaboration in safety

- Asks for and encourages input from people on issues that will affect them

- Helps others resolve safety-related problems for themselves

- Encourages others to implement their decisions and solutions for improving safety

- Seeks out and listens to diverse points of view

- Expresses confidence in the ability of others

- Supports the decisions that others make on their own

- Gains commitment of others before implementing changes

Figure 3-5. Behaviors that demonstrate Collaboration in safety.

4. Collaboration

Collaboration is an important aspect of great safety leadership. Essentially, it means "working together." Originally, the word "collaboration" was used to refer to scientists and others who work together in intellectual pursuits. In business and industry, it tends to mean "taking others' views into account before making decisions." The '80s ushered in an emphasis on employee involvement and participation, and since that time the word has gained strong currency in leadership generally.

Why is collaboration critical to effective safety leadership? Because safety involves all aspects of the organization, at all levels. Creating the right safety culture requires that every employee understand and buy into the core concepts and related behaviors and decisions that comprise safety excellence. This buy-in is more likely to occur at a more meaningful level if employees are part of the effort, and if they actually feel they are important to the effort. Participation and collaboration engender buy-in; independent decision-making shuts it down.

Communication

· Encourages people to give honest and complete information about safety, even if the information is unfavorable

· Keeps people informed about the "big picture" in safety

· Communicates frequently and effectively up, down, and across the organization

· Actively communicates and discusses safety information with direct reports

· Shares with people the background and reasons for safety policies and procedures

· Listens actively and with respect to safety concerns that are raised

· Constructively says what he or she is thinking

· Asks what others are thinking

· Makes sure that others feel comfortable and safe in raising issues and concerns

---

Figure 3-6. Behaviors that demonstrate Communication in safety.

5. Communication

Of course, communication is vitally important to effective safety leadership. However, it is one thing to say it, and quite another thing to do it. This applies to all levels of leadership in the organization. Despite knowing its importance, our performance and cultures still suffer from the lack of effective communication.

To make matters worse, we have grown accustomed to poor communication and tacitly accept it as a given. The effective safety leader must raise the bar on the necessity of communications excellence in safety. This is crucial to safety competence because it facilitates buy-in and participation, and because the very nature of safety effectiveness means each employee must understand what the safety issues are and how they are being addressed.

Recognition and Feedback

· Publicly recognizes the contributions of others

· Readily recognizes people for safety work well done

· Praises safety efforts more often than criticizes them

· Gives positive feedback and recognition for good performance

· Finds ways to celebrate accomplishments in safety

Figure 3-7. Behaviors that demonstrate Recognition and Feedback in safety.

6. Recognition and Feedback

Recognition and feedback do not refer to safety incentive schemes. We have written about the inadvisability of these approaches elsewhere[2]. Our experience indicates that safety incentive schemes usually have negative effects on organizational culture rather than positive ones.

The core principle of recognition and feedback is that performance improves when leadership notices positive change and responds to it. This response need not be formal or financial, but it must be consistent, especially when new behaviors start to emerge and need to be reinforced in order to become an established part of the culture.

The great safety leader is tuned in to the behaviors of subordinates and the larger organization, sets the expectation that behaviors and practices will occur, monitors regularly, and provides **soon-certain-positive** (Chapter 5) feedback when they do occur. Negative feedback also has its place in certain situations. But in the great majority of cases, soon-certain-positive feedback is most effective.

---

[2] Krause, T. R. and R. J. McCorquodale. 1996. "Transitioning Away from Incentives." *Professional Safety.*

Accountability

- Gives people a fair appraisal of their efforts and results in safety

- Clearly communicates people's roles in safety

- Fosters a sense that people are responsible for the level of safety in their organizational unit

- Sets clear responsibilities in safety for direct reports

- Holds people accountable for meeting their commitments

- Regularly reviews with direct reports indicators of their safety performance

---

Figure 3-8. Behaviors that demonstrate Accountability in safety.

7. Accountability

This is the last practice on the list and its effectiveness is based on having the other practices in place and working well. However, in most organizations we have consulted with this is the attribute of safety leadership that is most strongly in place. The danger in this situation is that naked accountability breeds an environment of resentment and distrust. Employees know they will be held accountable, but not given the resources, information, leadership, support, and encouragement they need to accomplish the task.

On the other hand, when the six practices that precede accountability are in place, this last practice is much easier to understand and has a higher degree of effectiveness. Much has been written about accountability elsewhere,[3] so we will only comment briefly. In the safety area, it matters what employees are held accountable for. Holding employees accountable for incident frequency rates *only makes sense if the numbers are statistically valid.* Activities that produce results should be measured and accountabilities established around them.

[3] Kraines, G. A, M.D. 2001. *Accountability Leadership: How to Strengthen Productivity through Sound Managerial Leadership.* Franklin Lakes: Career Press.

Measuring Safety Leadership Best Practices
(And Leadership Style)

We have found it most effective to measure leadership style and best practices by asking people around the leader how he or she is perceived. There are various tools to do this. We have used the Leadership Diagnostic Instrument (LDI) and our clients have found it useful for both individual and group feedback. If leadership coaching is desired, the LDI can be used to set the stage for it.

Results are shown in percentiles so each leader can compare his scores with those in a large database of other leaders' scores. This leads naturally to the development of an action plan for the individual safety leader. Based on the feedback found in the diagnostic instrument, what behaviors need to be changed?

A sample report is shown in Figure 3-9.

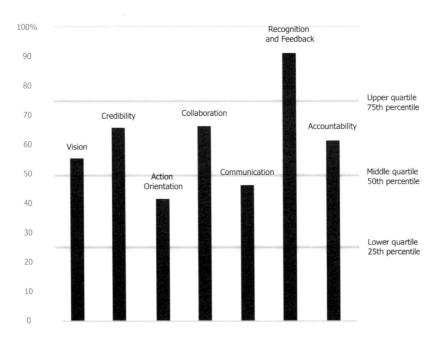

Figure 3-9. Sample Leadership Diagnostic Instrument report showing best practices scores expressed as percentiles.

Looking Forward

We've now considered the key elements of effective safety leadership, starting with the personality, values, and emotional commitment of the leader, and moving to leadership style and best practices. Next, we will consider the kind of organizational culture that is created when these leadership qualities are present.

• • • • •

Chapter 4

The Safety Leadership Model, Part 3
Understanding Organizational Culture
and Safety Climate

This chapter was written in collaboration
with Kim C. M. Sloat, Ph.D.

- Primary dimensions of organizational culture and safety climate

- Why some organizations respond to change more readily

- Measuring organizational culture and safety climate:
  the Organizational Culture Diagonostic Instrument (OCDI)

- The Organization Dimension

- The Team Dimension

- The Safety-Specific Dimension

So far this book has looked at three critical aspects of effective safety leadership: 1) the personality, values, and emotional commitment of the leader; 2) leadership style; and 3) best practices. The remaining outer ring of the Safety Leadership Model concerns organizational culture and safety climate, the subjects of this chapter.

We hear the word "culture" frequently in reference to organizational life. "He doesn't understand our culture." "Improving that part of performance will take a change in culture." But using the word and really understanding its implications are two different things. In our experience working with organizations on safety improvement, *leaders often fail to understand the implications of the decisions they make on the culture of their organization.*

Figure 4-1. The nine dimensions of organizational functioning within the Safety Leadership Model

Organizational culture can be difficult to understand because it is hidden. The values that drive the organization are frequently unstated, taken for granted. In one sense, everyone knows these things, but in another, we lose sight of them. And this lack of visibility can have serious effects.

Let's use an example. Suppose our organization's culture has a value for "getting the job done." Our CEO is known as a driver, willing to give resources to meet objectives, but expecting performance results in every case. As a result, each business unit has aggressive objectives and each business unit leader drives the organization for results. How do these dimensions influence our culture and what does that have to do with safety performance?

Driving hard to meet objectives is a good thing. The senior leader who sets objectives *intends* that certain standards and practices will be upheld in the process of meeting objectives, and that safety will not be compromised in the process. But by the time this gets to the plant level, it may create a cultural value that is so strong for production that in fact it does compromise safety. More than one organization we have worked with has learned this the hard way.

Many investigations into a fatality have found that decisions based on short-term production objectives had set the tone for compromising safety. This is not to say that safety and productivity are a trade-off; in fact, the safest plants are usually also the most productive. But it does suggest that leaders can create effects on culture without even knowing they are doing it.

Defining the Primary Dimensions of Organizational Culture and Safety Climate

After reviewing more than fifty studies in the research literature, we isolated nine dimensions that define organizational culture and safety climate. (Table 4-1). Each dimension has been shown to predict safety outcomes.[1]

---

[1] Hofmann, D.A. 1999. *A Review of Recent Safety Literature and the Development of a Model for Behavior Safety*. Ojai, CA: Behavioral Science Technology, Inc.

Why Do Some Organizations
Respond to Change More Readily Than Others?

As pressure for improved organizational performance accelerates, employees are being asked to go beyond their traditional job duties and to take more responsibility for their work. In some organizations, employees are easily engaged, rise to the challenge, and even give discretionary time to assure that goals are met. In others, the need for change meets resistance, and employees are unwilling to extend themselves. What determines the employees' response to the need for change? Why do some organizations adapt easily and others struggle? Answering this question adequately is important to improved safety, since many organizations will find it necessary to bring about fundamental "culture change" to reach safety excellence.

The nine dimensions of organizational culture are also important to understanding organizational change. Change efforts do not occur in a vacuum. Organizational members usually have long histories with each other. An individual manager may have had thousands of interactions with his or her manager and with peers. A series of superiors has probably come and gone, and each one probably focused on particular areas and neglected others. These interactions teach managers and workers what is important to others in the organization, how they are likely to be treated in various circumstances, and whether others are likely to do what they say they will do.

An individual's experience with the organization congeals into a set of perceptions, or beliefs, about the way things are. These beliefs influence how he behaves, and they define the organization's culture and safety climate. We can learn about an organization's culture and climate by asking its members questions about how they perceive various aspects of organizational life. These questions can be grouped into scales or dimensions, as shown in Table 4-1.

Perhaps surprisingly, *the success of most change efforts depends more on perceptions about some basic aspects of organizational life* than on perceptions specific to the area to be changed. For instance, improvements in safety at the front-line level depend more on workers' perceptions of how they are treated by their supervisor than on perceptions of the importance of safety

1	Procedural Justice	The extent to which the individual worker perceives fairness in the supervisor's decision-making process.
2	Leader-Member Exchange	The relationship the employee has with his or her supervisor. In particular, this scale measures the employee's level of confidence that his supervisor will go to bat for him and look out for his interests.
3	Management Credibility	A perception of the employee that what management says is consistent with what management does.
4	Perceived Organizational Support	The perception of the employee that the organization cares about him, values him, and supports him.
5	Workgroup Relations	The perception the employee has of his relationship with co-workers. How well do they get along? To what degree do they treat each other with respect, listen to each other's ideas, help one another out, and follow through on commitments made?
6	Teamwork	The extent to which the employee perceives that working with team members is an effective way to get things done.
7	Safety Climate	Scales 7, 8 and 9 are specific to safety performance. The Safety Climate scale measures the extent to which the employee perceives the organization has a value for safety performance improvement.
8	Upward Communication	The extent to which communication about safety flows freely upward through the organization.
9	Approaching Others	The extent to which employees feel free to speak to one another about safety concerns.

Table 4-1. Nine dimensions of organizational functioning.

in the organization. Similarly, perceptions about safety at the front-line level define certain aspects of the culture.

Most importantly, improvement efforts are more successful when perceptions in key areas are well understood. It is then possible to build upon favorable perceptions and undertake targeted improvement where perceptions are unfavorable.

In most situations the front-line workers are at the greatest risk for injury. Therefore, it is important to find out what perceptions influence workers' safety-related actions.

The nine dimensions identified in Table 4-1 include workers' perceptions about supervisors, co-workers, managers, and the organization as a whole. It's important to note that six of these areas are not specific to safety. It may seem odd that safety outcomes would be strongly influenced by dimensions that do not appear to have much to do with safety. However, as discussed previously, those companies that achieve excellence in one area of performance tend to achieve it in many others. High-performing organizations tend to be good at everything. If the general environment in an organization is favorable, safety initiatives will tend to be successful; if the environment is less favorable, the initiatives will be less successful.

The arrows in the Figure 4-2 show the direction of influence. Each relationship shown is statistically significant and wider lines on the arrows denote stronger predictive relationships.

The scales of the Organization Dimension in Figure 4-2 are the most fundamental — perceptions about these scales influence safety outcomes directly and indirectly through perceptions about team functioning and dimensions related specifically to safety. Notice that the scales in the Organization and Team Dimensions are not safety-specific. This leads to two important points. First, for the Safety–Specific Dimension we could substitute equivalent dimensions related to another area of organizational functioning, such as reliability or quality. We would likely find the same pattern of influences on these outcomes as is true for safety. Second, long-term improvement in safety is unlikely without attention to organizational and team variables. An organization with poor relations between workers and supervisors, and dysfunctional workgroups, will find it difficult to sustain gains that come from change efforts focused too tightly on safety alone. Long-term high performance in safety is much more likely if perceptions in these key areas are favorable.

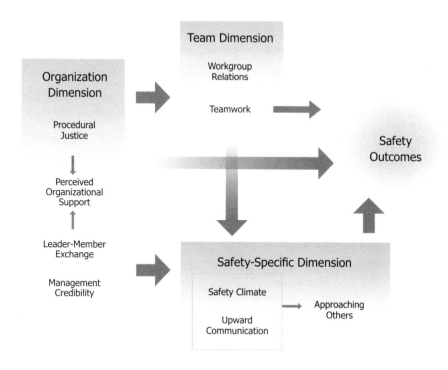

Figure 4-2. How organizational functioning dimensions relate to safety outcomes.

Who or what the perceptions refer to	FOCUS OF THE PERCEPTIONS	
	General	Safety Specific
First-line supervisor	Fairness of decision-making processes (Procedural Justice) Strength of working relationship (Leader-Member Exchange)	Climate around bringing up safety concerns (Upward Communication)
Company/Management	Honesty and consistency of managers (Management Credibility) Organizational concern for employees' needs and interests (Perceived Organizational Support)	Value of and priority for safety (Safety Climate)
Co-workers	Team effectiveness (Teamwork) How well group members get along (Workgroup relations)	Likelihood that workers will speak to others about safety (Approaching Others)

Table 4-2. Perceptions that influence safety outcomes organized by target of the perception and whether the perception is general or safety-specific.

Measuring Organizational Culture and Safety Climate: The Organizational Culture Diagnostic Instrument (OCDI)

The first step to improvement is to measure perceptions. This is ordinarily done using a diagnostic instrument that measures perceptions and compares them to those from other organizations. It is then possible to develop an intervention plan to make improvements.

It's worth noting that diagnostic instruments and surveys are not always the same thing. A survey is a set of questions thought to be important. We have employees answer them, roll up the data, and then examine it. Often employees complain of "being surveyed to death." Then, after going to all the trouble of taking the survey and looking at the data, not much seems to change. As a result, the credibility of leadership declines. Many organizations need to do fewer surveys and take more action based on valid findings.

A diagnostic instrument is a survey that has been shown to predict specific outcomes and has a normative database so that one organization can be compared to another. Users can pinpoint low-scoring areas of functioning and design improvements that will impact organizational performance.

Diagnostic Instrument	**Survey**
Has established predictive relationships between survey scales and performance outcomes	Set of questions thought to be important
Provides scores in relation to a database for comparison across organizations	
Pinpoints specific areas for improvement	

Figure 4-3. Diagnostic instrument vs. surveys.

In the rest of this chapter, we will discuss in more detail each of the dimensions that influence safety outcomes.

Organization Dimension

The ideas underlying the four dimensions are not specific to safety but have to do with organizational functioning considered broadly:

- Procedural Justice (fairness of decision-making)
- Leader-Member Exchange (supervisor-worker relationships)
- Management Credibility (honesty, consistency, and competence)
- Perceived Organizational Support (organizational concern for needs and interests of employees).

These four dimensions can be considered the "four pillars of culture." When leaders say they need to change the organization's safety culture they are likely referring to some aspect of these four dimensions.

The Principle of Reciprocity

These critical foundations of excellent organizational performance can be understood from social exchange theory.[2] This theory says that important aspects of relationships (between individuals, or between an individual and a group) can be viewed as a series of exchanges or interactions in which the principle of reciprocity plays a central role. For example, if an employee is treated with dignity and respect and offered support by his or her supervisor, the likelihood increases that the employee will reciprocate; job performance, extra-role behavior, and loyalty will tend to increase. On the other hand, if the worker feels demeaned or disrespected, he is much less likely to fully engage in the work.

The three dimensions based on exchange theory tap different relationships: with the supervisor (Leader-Member Exchange), with managers (Management

---

[2] Greenberg, J. & R. Cropanzano (Eds.), *Advances in Organizational Justice*. Stanford, CA: Stanford University Press.

Credibility), and with the organization (Perceived Organizational Support). Of these, the relationship with the supervisor is the most important. The quality of that relationship strongly influences whether the employee believes the organization cares about his needs and interests (Perceived Organizational Support). An employee who has a great relationship with his supervisor also tends to score high on Perceived Organizational Support. To a large extent, the supervisor is the embodiment of the organization for the employee. Each of these dimensions predicts safety performance outcomes independently; together they are even more powerful.

The scales underlying the Organizational Dimension are important drivers of most of the other dimensions. For instance, team functioning is strongly influenced by the aspects of organizational life represented by the Organizational Dimension. This makes sense — if a supervisor treats people fairly and has good relationships with them, the overall team should function more effectively. Employee beliefs with respect to the Organizational Dimension scales also influence perceptions around safety, the Safety-Specific Dimension. This also makes sense. For instance, one of the scales in the Safety Dimension is Upward Communication (employees raising safety concerns with the supervisor). Good relationships and fair treatment are likely to create a favorable climate for bringing up such issues. The Organization Dimension also directly influences safety outcomes (safe behavior, injury, injury reporting). If managers are seen as fair, consistent, and competent, employees are more likely to report injuries.

Procedural Justice

Perceptions of procedural justice are powerful influences in an organization. Fair procedures are characterized by:

- **Consistency**: Across persons and time

- **Lack of bias**: Avoidance of personal self-interest by decision maker(s)

- **Accuracy**: Decisions are based on good information and informed opinion

- **Correctable**: One can appeal decisions made at various points of the process

- **Representativeness (or "voice")**: The procedure reflects the basic concerns, values, and outlook of those affected

- **Ethicality**: The procedure is compatible with the fundamental moral and ethical values of those affected

If leaders seem to be making decisions in fair ways, workers assume that they can obey and follow rules without worrying too much about exploitation or rejection. How a company handles employee safety concerns is one specific safety-related situation. If employees consider the procedures to be fair, they are likely to be more accepting of an unfavorable outcome on a specific issue. They are also more likely to believe that over the long run important issues will be addressed. Other factors include how workers are treated following an injury and the manner in which safety rules and procedures are developed. Procedural Justice also affects organizational effectiveness in indirect ways. Employees interpret those procedures that treat them with dignity and respect as being fair.

Leader-Member Exchange

The leader-member exchange concept developed from attempts to understand exactly how leaders influence subordinates. That is, what can leaders do to get desired performance from employees? Chapter 2 described one method:

positive consequences for meeting or exceeding expectations, and negative consequences for failing to do so. This kind of arrangement is called transactional leadership (or sometimes "management"). Supervisors, formally or informally, strike a deal (have a transaction) with workers: "You do 'x', and 'y' will happen to (or for) you." Another approach is called transformational leadership (or sometimes just "leadership"). With this arrangement, the worker tries to achieve a goal not because of an explicit reward or punishment, but because he sees achieving the goal as fulfilling an organizational purpose important to the leader *and* to himself.

Transformational leadership exerts influence principally through relationships with employees. In a workgroup, the supervisor develops relationships with each of the workers. The leader exerts influence by getting each person to see how his or her objectives support the larger objectives of the organization.

Management Credibility

Management Credibility can be viewed as an attitude held by one person towards another, based on the first person's observations of the other's behavior. Most research in this area has focused on the subordinate's trust in the manager and has focused on manager behaviors that lead to perceptions of trustworthiness. Various perspectives have been used to understand the development of trust. One of these is social exchange theory. From this point of view, managers initiate the development of trust by acting in ways that provide benefits to followers (e.g., reducing uncertainty). Over time, the odds that followers will trust the manager will increase, and they will behave in ways that provide benefits to the manager (e.g., cooperation).

Perceptions that a manager is competent seem to be a necessary but not sufficient basis for development of trust. That is, workers are unlikely to trust a manager who is seen as incompetent, but competence alone does not necessarily lead to trustworthiness.

Manager behaviors that influence perceptions of trustworthiness include:

1. **Consistency:** Reliability or predictability over time and in various situations.

2. **Integrity:** Consistency between word and deed, including
   - Telling the truth
   - Keeping promises

3. **Sharing control**: Participation in decision-making and delegation

4. **Communication**
   - Accurate, forthcoming information
   - Explanations of decisions and timely feedback on them
   - Open exchange of thoughts and ideas

5. **Demonstration of concern (benevolence)**
   - Consideration and sensitivity of employees' needs and interests
   - Acting in a way that protects employees' interests
   - Refraining from exploiting others for one's own benefit

Perceived Organizational Support

Why would an employee go the extra mile for the organization by, for example, being on a safety committee? A strong influence is whether the employee believes the organization is concerned with his needs and interests. That is, does he perceive that he is supported by the organization? Perceived Organizational Support (POS) can be understood through social exchange theory, as can Leader-Member Exchange (LMX) and Management Credibility. In the case of LMX, the exchange is between the employee and supervisor; with POS, it is between the employee and the organization. The central notion in social exchange is reciprocity, responding in kind to how one is treated. With respect to POS, it is important that the favorable treatment is seen as discretionary on the part of the organization. That is, if a certain benefit or procedure is required by law or contract, employees will generally not see it as evidence of caring and concern by the organization. If employees believe the organization cares about and extends itself for them, they are more likely to extend themselves for the organization.

Suppose a company strives for excellence in safety and goes well beyond what is required to produce an injury-free culture. Freedom from risk of personal injury is an emotional issue, and companies that strive for excellence in safety communicate their concern for employees. That concern is likely to be reciprocated by extra effort by employees in the areas important to the company.

POS is not the same as job satisfaction, although the two are often related. Employees who believe the organization cares about them are more likely to be satisfied. POS is an overall perception by employees of organizational commitment to them, whereas job satisfaction is an affective (positive/negative) response to specific aspects of the work situation (e.g., pay, physical working conditions, work schedules).

The strength of the relationship with the immediate supervisor (LMX) affects worker perceptions of POS. It is as if an employee sees his relationship with the supervisor as representing the organization's concern for him. More generally, employee perceptions of the extent to which managers, supervisors, and (to a lesser extent) co-workers are trustworthy and supportive affect POS.

**Organization Dimension Scales:
Procedural Justice, Leader-Member Exchange,
Management Credibility, and Perceived Organizational Support**

Unfavorable Perceptions	Favorable Perceptions
• Taking hostile actions against co-workers	• Organizational citizenship behavior (going above and beyond the call of duty, such as volunteering for safety roles)
• Formal relations between supervisor and worker	• Positive perceptions of the organization's value for safety
• Few opportunities for worker input	• Commitment to the organization
• Low alignment between supervisor and worker goals	• Less resistance to change
• Unwillingness by workers to go beyond formal job requirements	• Workgroup functioning well as a team
• Worker intentions to leave the organization	• Worker willingness to raise safety concerns
• Disengagement — low commitment to the organization	• Mutual trust, respect, influence and obligation between supervisor and worker
• Low levels of initiative and risk-taking	• Empowerment of workers by supervisor
• Absenteeism	• Supervisor encouragement of initiative by workers
	• Higher levels of performance by workgroup
	• Willingness of workers to expend extra effort
	• Overall job satisfaction
	• Satisfaction with supervisor
	• Quality of communication between managers and reports
	• Free exchange of information and knowledge within the organization
	• Organizational performance
	• Cooperation and teamwork
	• Willingness of individuals to seek help when needed
	• Trust

Table 4-3. How Organization Dimension scales are manifested in the workplace.

Team Dimension

There are two aspects to team functioning — how effectively the team gets work done (Teamwork) and how well co-workers get along (Workgroup Relations). Perceptions of these aspects are highly related but they represent distinct dimensions. For instance, a workgroup could be unproductive but have members who get along well with each other.

Perceptions of team functioning (Team Dimension) are affected by perceptions of more fundamental issues in the organization (Organization Dimension). An organization that has fair procedures, good relations between workers and supervisors, trustworthy managers, and concern for employees will tend to have well-functioning teams. Not surprisingly, basic aspects of how employees are treated set the stage for team effectiveness and cohesion.

The Team Dimensions have direct effects on safety outcomes: level of safe behavior, injuries, and injury reporting. Team functioning also affects perceptions about the organization's value for safety, the climate around raising safety issues, and the likelihood of workers talking to one another about safety-related behavior. These perceptions in turn affect safety outcomes so that teamwork and work group relations have both direct and indirect effects on safety outcomes.

In addition, the quality of relations within a team influences the climate in the team for change. Higher functioning teams are more open to change.

Teamwork

Various dimensions affect team functioning, including design of the work and the team (socio-technical considerations), team composition, the general organizational context in which the team operates, and internal group processes. The teamwork dimension represents an overall assessment of group cohesiveness and functioning, the result of the various influences.

Teamwork is affected by fair treatment of the members, both by the supervisor and by peers. If both the supervisor and the group itself make decisions by processes that are considered fair, group members will have positive attitudes towards the supervisor (trust) and the group (commitment). This leads to better team functioning.

Team Dimension Scales: Teamwork and Workgroup Relations	
Unfavorable Perceptions	**Favorable Perceptions**
• Hostile actions between group members • Reluctance to take risks interpersonally • Higher turnover • Resistance to authority	• Talking to one another about reducing exposure to hazards • Raising safety concerns with the supervisor • Higher levels of safety involvement • Team member satisfaction with co-workers, the work, and supervisor • Higher team performance • Greater likelihood of helping out co-workers • Higher commitment to the group • Fewer accidents

Table 4-4. How Team Dimension scales are manifested in the workplace.

Workgroup Relations

Social relationships within the workgroup influence important safety-related variables. This is fairly easy to picture. In a group in which people do not get along well together, individuals are less likely to go out of their way to speak up to co-workers about safety. Speaking up can be risky — one cannot be certain how the other person will react. Likewise, raising a safety concern in a safety meeting is risky — other group members might ridicule the concern. When there are low levels of trust, workers are less willing to take these risks. On the other hand, if relations between group members are good, people will feel more comfortable interacting around safety issues and raising concerns.

Trust is related to how well the team functions (teamwork). In high-performing teams, members are more likely to identify with the team. Identification leads to trust among team members, which results in cooperation. Dysfunctional groups with a low sense of team identity will have low levels of trust. Social relationships among group members are a strong predictor of worker compliance with safety rules and procedures.

Workgroup relations are affected by the leader of the group. Supportive and trustworthy behavior by the leader is likely to lead to trust among members of the group.

Safety-Specific Dimension

The scales in this dimension represent three different links between organizational and team functioning, and safety outcomes such as injuries. One link is through workers raising safety concerns (Upward Communication). A second is through workers speaking to one another about exposure to hazards (Approaching Others). A third link is individuals taking responsibility for their own safety. This sense of responsibility is strongly influenced by perceptions of the Safety Climate, which also influences Upward Communication and Approaching Others. Relations with the supervisor and co-workers, and a sense of fair treatment by the organization and the supervisor affect whether workers will raise concerns to the supervisor or to co-workers.

Safety Climate

The idea of a Safety Climate gained attention around 1980. The distinction between safety climate and culture is somewhat controversial. (Figure 4-4) Generally, culture is seen as the "background" and climate as the "foreground." Culture is a more fundamental concept, with climate and culture influencing each other. Culture has been variously defined as "the way we do things around here," and "shared common values." Climate is generally considered to be the climate *for* something: safety, quality, service, etc. It is employee perceptions about what gets rewarded, supported, and expected in a particular setting. One can talk about *the culture of an organization*, but not *the climate of an organization*; rather, it would be the climate for a specific performance indicator (e.g., safety, reliability, cost, etc.) in the organization. Climate is more readily changed than culture.

Specific measures of safety climate vary, but a common underlying theme is management commitment to safety. Across a number of research studies, scores on this dimension have been related to safety outcomes such as injury rates. The underlying logic is that a high level of commitment to safety

Culture and Climate	
Culture	**Climate**
Common values that drive organizational performance	Perceptions of what is expected, rewarded, and supported
Applies to many areas of functioning	Applies to a specific area of functioning
"How we do things"	"What we pay attention to"
Unstated	Stated
Background	Foreground
Changes more slowly	Changes more rapidly

Figure 4-4. Comparing culture and climate.

would result in visible support in the form of resources and programs, for example. This support would result in positive perceptions of organizational commitment, which would influence how people work on a day-to-day basis. The various safety processes would also reduce hazards, leading to lower injury rates.

Perceptions about Safety Climate are influenced by perceptions of the dimensions represented by the Organization Dimension. In particular, there is a strong relationship between Perceived Organizational Support and Safety Climate. Workers who believe the organization cares about them in general are also likely to believe that management is committed to safety. Commitment to safety is one specific way in which organizational support can be demonstrated.

Upward Communication

Another link between organizational and team variables and safety outcomes is whether workers raise safety concerns. For instance, workgroups characterized by supervisor fairness and support have fewer injuries. How does fairness and support result in better safety outcomes? One mechanism is that workers speak up about safety concerns. A supervisor who is fair and supportive is more likely to listen to concerns, and respond appropriately. Over time, the willingness of workers to identify opportunities for improvement, and the supervisor's commitment to get action, will reduce exposure to hazards, and thus reduce injuries.

Perceptions of Upward Communication are related to scores on Perceived Organizational Support and Leader-Member Exchange. Workers who have good relationships with their supervisor, and believe the organization cares about them, are more likely to bring up safety concerns. Team functioning and relations also affect the willingness of workers to raise safety issues. In a dysfunctional team, workers will be more reluctant to bring up issues that might elicit negative reactions from co-workers. If the supervisor is open to upward communication about safety issues, that sends a strong signal to workers that the organization values safety.

Approaching Others

The Upward Communication dimension deals with workers raising safety issues with the supervisor — typically, these issues involve facility or equip-ment items, and perhaps procedures. Another opportunity for improvement lies with involvement. In a healthy safety climate, workers will speak up to one another about ways to reduce exposure. The more that co-workers pay attention to exposure, the safer the Working Interface.

Approaching Others is related to both Leader-Member Exchange (LMX) and the commitment of the team leader (supervisor) to safety. The quality of the relationship with the supervisor is related to the willingness of team members to speak up. If the leader values safety, the subordinate can recip-rocate high-quality LMX by speaking to others about safety.

Team functioning also influences Approaching Others. In a high-function-ing team with good interpersonal relationships, members will be willing to speak up to one another, confident of getting a reasonable reaction from their co-workers. By contrast, in dysfunctional groups, reactions from co-workers will be unpredictable, or predictably negative.

Safety-Specific Dimension Scales: Safety Climate, Upward Communication, and Approaching Others	
Unfavorable Perceptions	**Favorable Perceptions**
• Workers are more likely to attribute the cause of an accident to situational elements even when worker behavior was a major factor	• Higher levels of involvement and initiative • Lower injury rates • Higher levels of injury reporting • Individuals feel more responsible for their own safety and that of others • Higher individual commitment to safety • Greater likelihood that workers will raise safety concerns

Table 4-5. How Safety-Specific Dimension scales are manifested in the workplace.

Summary

Research into the organizational influences on safety outcomes has confirmed some long-held beliefs, and turned up some surprises. Over the years, much has been written about the importance of management commitment to safety. The link between strong commitment to safety and good outcomes is now well established. Further, it is clear what commitment means: managers who are knowledgeable about safety in their area of responsibility, who assign a high priority to safety in arenas such as management meetings, and who take action and commit resources to improve safety. Two other findings, though not surprising, are more specific than most thought: safety outcomes are better in organizations in which first-line supervisors encourage employees to bring up safety concerns and in which workers take the initiative to talk to one another about safety.

The surprises relate to aspects of organizations that don't directly concern safety. Safety results are better in workgroups in which members have good working relationships with the supervisor, in which the supervisor is fair in making decisions, and in which the team is productive and members get along well with each other. In other words, workgroups that function well in general also do well in safety.

Another surprise is that if workers believe that managers are trustworthy, honest, and consistent in general, they have fewer injuries (and are more willing to report the ones they do have). Further, if employees believe the organization cares about their concerns and interests, they are more willing to extend themselves on behalf of the organization. If safety is valued in the organization, workers will expend extra effort to work safely and to improve safety.

• • • • •

The Leader's Role:
Understanding Two Crucial Aspects
of Human Performance

Improving safety is a significant organizational change. But organizations don't change just because leaders want them to do so, they change when leaders change their own behaviors. That may seem to be an overwhelming obstacle. Often the leader knows certain behaviors should change, but finds it difficult to understand how to change them. Applied behavior analysis provides a method for understanding and changing behavior. But although behavior change is critically important to improving safety, it isn't the entire answer. The improvement process also demands that leaders use their cognitive skills to make crucial decisions. Safety leaders need to understand the potential detrimental effects of cognitive bias on the decision-making process.

Chapter 5

Changing Behavior
Using Applied Behavior Analysis

- Applied behavior analysis in organizational settings

- How applied behavior analysis supports safety improvement

- Central concepts: Antecedents, behavior, and consequences

- ABC analysis as a tool

- Example: Changing behavior at the leadership level

- Considerations for identifying new consequences

- Example: Changing behavior at the middle-management level

- Putting behavior analysis to work

In practically all organizations, a gap exists between the way the organization intends for things to be done, and the way things are actually done. This is so much a fact of organizational life that most of us have grown used to it and more or less accept it as an inevitable fact.

None of this should come as a surprise, since we also know that our own behavior, and that of those around us, is also different than what we intend or desire. We know we'd like to eat certain diets, engage in certain particular routines, get important errands done, and that we fall short of doing what we intend or want to do. We are frustrated and perplexed, sometimes by our own behaviors and often by the behaviors of those around us.

Generally, the behavioral sciences offer us little real help in understanding and changing these things. There are many theories, but they don't really get to the point of reliability, and we remain perplexed.

Applied behavior analysis is a tool that sheds light on these difficulties.

Applied Behavior Analysis in Organizational Settings

Applied behavior analysis (ABA) is a powerful methodology for understanding, measuring, and influencing behaviors of all kinds. The application of ABA to organizational performance has been written about extensively elsewhere,[1] so we will present only a summary in this chapter. Since there is some controversy about the use of ABA, we will also comment on where it is best used, and how to avoid misusing it. Most importantly, we will discuss the most efficient and effective use of ABA for helping the organization improve safety performance.

Behavior analysis is a methodology drawn from the academic discipline of psychology. It is uniquely well grounded in empirical research studies that span a period of about fifty years. The method comes originally from the work of Harvard psychologist B.F. Skinner, a controversial researcher whose impact on how we view psychology and human behavior has been immense. Unfortunately, early work in the application of behavior analysis

---

[1] Komaki, J. 1986. "Applied Behavior Analysis and Organizational Behavior: Reciprocal Influence of the Two Fields." *Research in Organizational Behavior.* 8: 297-334.
Krause, T.R. 1997. *The Behavior-Based Safety Process: Managing Involvement for an Injury-Free Culture, 2nd Ed.* New York: Van Nostrand Reinhold.

to industrial/organizational settings, while startlingly successful, was plagued by misunderstandings, the effects of which remain today.

Research in education, clinical psychology, and organizational improvement showed that the application of behavior analysis was uniquely useful, primarily because it nailed down concrete measures of actual performance, in observable terms. Today, we take for granted the need to specify the actual behaviors we want to influence when organizational change is needed. Prior to the work of Skinner, this was unheard of. Vague statements such as "improving the attitudes of managers" were commonplace. At the same time, the unfortunate term "behavior modification" arose and was misunderstood to be a kind of sinister manipulation. The popular press and parts of academia created a set of misconceptions; behavior analysis was a device of "Big Brother," who would be watching and insidiously changing the way you act, "modifying your behavior." So, despite the birth of a powerful method for influencing behavior, the popular view of it was based on a set of misunderstandings.

In the organizational performance change arena, early studies were very promising. However, neither consultants nor industrial leaders adequately understood what was required to implement the method properly. Consultants from the academic world understood the technical aspects of ABA, but did not understand organizational realities: the unstated ways that organizations outside the academic world function. And organizational leaders didn't recognize the depth of value and complexity contained in a method that would actually measure and change behavior. The result was that while many excellent studies were published showing the effectiveness of organizational behavior change, the method has never been fully embraced within industry.

How Applied Behavior Analysis Supports Safety Improvement

In spite of these problems, industry has taken notice of the fact that *getting to the behavioral level* matters to performance outcomes. Whatever you may think of "behaviorism," organizational leaders universally recognize that performance improvement comes down to behavior change. Whether at the level of the CEO, the supervisor, or the front-line employee, *if the organization is to improve, its performance behaviors have to change. And if the culture is to change, behaviors that influence and support the culture must change.*

This applies to safety improvement just as it applies to every other kind of organizational change. It applies to all levels of employees, from the senior-most leader to the most recently hired front-line employee. The task is to identify and systematically improve those behaviors at each level in the organization that reduce employee exposure to hazards. To assure sustained safety improvement, leadership must take on this task vigorously.

Of course, it is one thing to talk about the need for behavior change and quite another to assure that it happens. We all know that a gap exists between what is said and what is done. Choose any existing procedure in your organization and then observe how the activity described by the procedure is actually done. You will almost always find serious differences. The easy part is determining how things should be done; the hard part is leading in such a way as to assure that they actually happen.

Achieving reliable execution is a challenge for most of the companies we work with. What looks good on paper often doesn't translate into action once faced with the cold light of daily demands and priorities. All of this points to and emphasizes the critical importance of being able to understand and influence behavior effectively. Applied behavior analysis is a uniquely effective tool for addressing execution.

Central Concepts:
Antecedents, Behavior, and Consequences

Applied behavior analysis begins with understanding the three basic concepts: the antecedent, the consequence, and the behavior itself. An *antecedent* is an event that precedes and triggers a behavior. *Behaviors* are simply observable acts. A *consequence* is any event that follows a behavior. A simple example is the ringing doorbell (antecedent), which we answer (behavior) to see who is at the door (consequence). Common sense tends to identify the antecedent, in this case the doorbell, as the cause of the behavior, in this case answering the door. And of course, the antecedent is important. However, it will turn out that while both are influential, consequences are more powerful determinants of behavior than antecedents.

Suppose there's a situation in which the doorbell rings repeatedly and there is no one there. Perhaps the bell is malfunctioning, or pranksters are ringing the bell and running away. In such a case, the behavior of answering the door to see who is there is frustrated by lack of the expected consequence. In fairly short order, one would stop "automatically" answering the door. As soon as the ringing doorbell no longer reliably signals the presence of a caller at the door, it no longer elicits the behavior of going to the door to see who is there. By itself, the antecedent (the doorbell) does not directly determine the behavior (answering the door). Instead, antecedents elicit certain behaviors because they signal or predict consequences.

In a nutshell, behavior analysis involves the following principles:

- Both antecedents and consequences influence behavior, but they do so very differently;

- Consequences influence behavior powerfully and directly, and

- Antecedents influence behavior indirectly, primarily serving to predict consequences.

Many well-intended change initiatives fail because they rely too much on antecedents — things that come before behavior — goals, best practices, meetings, and so on. Too often, these same antecedents have no powerful consequences to back them up.

In addition to discovering that consequences are stronger than antecedents, behavioral science research has found that in the competition of consequences to control behavior, some consequences are stronger than others.

"Soon/Certain/Positive" — The Strongest Consequence

There are three qualities that determine which consequences are most powerful:

- **Timing-** A consequence that follows soon after a behavior influences behavior more effectively than a consequence that occurs later.

- **Consistency-** A consequence that is certain to follow a behavior influences behavior more powerfully than an unpredictable or uncertain consequence.

- **Significance-** All things being equal, a positive consequence influences behavior more powerfully than a negative consequence.

Both positive and negative consequences influence behavior. Which is more powerful depends on several considerations. But for optimal use in organizational change, positive consequences are more effective for a number of reasons: their use creates a positive environment, behaviors changed by positive consequences tend to generalize more readily, and they lack negative side effects.

This means that the consequences with the most power to influence behavior are those that are soon, certain, and positive. A consequence that increases the likelihood of a behavior occurring in the future is called a *reinforcer*. When behavior is strengthened by consequences, it is *reinforced*. *Reinforcement* is a mechanism by which behaviors are acquired. Behavior analysis goes to great lengths to specify types of reinforcement, conditions favorable to it, and methods for selecting optimal types and schedules. See *Behavior Analysis for Lasting Change* by Beth Sulzer-Azaroff and R.G. Mayer, or *Behavioral Modification in Applied Settings* by Alan E. Kazdin for a more detailed description.

These concepts can be integrated and put to use with a tool called ABC Analysis. This tool is useful wherever we wish to understand and influence critical behaviors. For our purposes, that means behaviors that contribute to exposing employees to hazards.

ABC Analysis as a Tool

ABC Analysis has three steps:

Step 1. Identify the target behavior you wish to influence, and state it in the negative. List the antecedents and consequences for the target behavior and list the qualities of the consequences. This will give you an understanding of why the behavior occurs and a basis for forming a strategy to change it.

Step 2. In positive terms, state the target behavior you have just analyzed. List the antecedents and consequences for it, as well as the qualities of consequences.

Step 3. Draft the action plan showing what steps will be taken to assure that the right antecedents and consequences are delivered to influence the target behavior.

Example 1: Changing Behavior at the Leadership Level

We will use a couple examples to illustrate the use of this tool. Let's say we are working with a manufacturing vice president who is motivated to improve safety in his or her organization. This leader is motivated by the fact that of the eight locations under his supervision, performance varies widely, even though the locations operate the same technology and have similar workforces. This executive has been directed to reduce injury frequencies. While safety initiatives initially reduced these, no further reduction has occurred for almost two years. The site managers who report to the vice president are also motivated to improve, but are less clear about the necessity and *don't seem to spend adequate time finding new and innovative solutions.* Discussions with the vice president yield the following critical behavior of the plant managers: failure to spend adequate time on safety. Figure 5-1 shows the analysis. (We have provided a tool to assist with this type of analysis on the CD in the back of this book).

Step 1: Using the example above, identify the target behavior you wish to influence, and state it in the negative. List antecedents and consequences and specify the qualities of each consequence.

A	B	C	
Other priorities	Failure to spend adequate time on EHS activities	Injury	S/LC-
Putting out fires		Negative Feedback	S/LC-
Other people's job		Get other things done	SC+
No one else does		Avoid discomfort	SU+
Lack of training		Maintain illusion of good safety performance	SC+
Lack of focus			
Too far from day-to-day safety issues			
Fear of demands I can't meet			

Figure 5-1. Step 1, analyzing the undesired behavior.

The analysis is helpful in that it starts to show us what this behavior by plant managers is about. We see that a set of antecedents (e.g., "I have other priorities") triggers this behavior (failure to come up with new safety solutions). These antecedents will be helpful in Step 2 when we want to understand the factors that influence the desirable behavior. The consequences in column C (e.g., "I can get other things done, although the boss might notice the lack of safety innovations") tell us what is maintaining this behavior. It becomes apparent that the site manager is receiving a number of soon/certain/positive consequences for this behavior. Although there are also negative consequences, they tend to be uncertain. This is a pattern we see in analyzing safety-related behaviors generally; the short-term consequences outweigh the long-term ones. Of course, this leads to a strategy that is fatally flawed. At some point, a serious injury or fatality will occur, and then the leader will realize his mistake. However, in the shorter term, our vice president is up against a serious barrier.

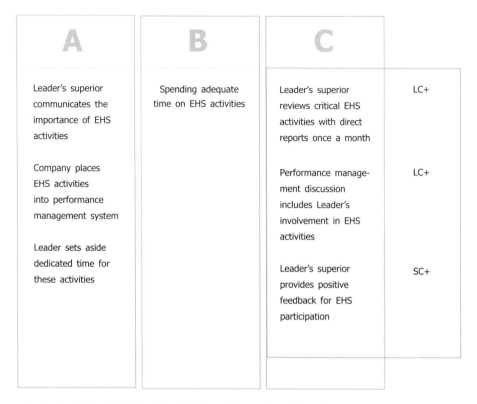

Figure 5-2. Step 2, analyzing the desired behavior with new antecedents and consequences.

Step 2 is to use the information contained in Step 1 to analyze the desired behavior. The analysis might look like Figure 5-2. Now we want to see what kinds of antecedents we can use to trigger the behavior we want. Examining the antecedents that trigger the opposite behavior is helpful. If "lack of knowledge" triggers the failure to perform safety activities, then providing training with safety content may be an effective antecedent to trigger doing safety activities.

Considerations for Identifying New Consequences

When the list of antecedents is complete, we ask ourselves: *What kinds of consequences can be provided to this site manager to encourage the behavior we would like to see?*

The answer to this question is controversial. Many would say that the consequences must be financial — the stronger the better. (This same line of thought would have us provide tangible consequences for front-line employees who perform well in safety.) Our experience leads to quite a different conclusion. "Safety incentives," tangible rewards made contingent on injury frequency rates, are often counter-productive. We have written about this extensively elsewhere[2], so we will just say here that we don't recommend their use, and that we have seen dramatic and documented improvements in safety with no material "incentives" whatsoever.

For safety improvements, our experience confirms that it is most desirable to use positive feedback as the primary soon/certain/positive consequence. To understand this in more depth, consider the performance improvement model shown in Figure 5-3.

This model is a hierarchy. The factors near the bottom are more fundamental than those at the top. And behaviors maintained by factors near

Performance Improvement

Figure 5-3. The Performance Improvement Model.

---

[2] Krause, TR 2000. "Motivating Employees for Safety Success." *Professional Safety.* 45: 22-26
Krause, T.R. 2001. "Moving to the Second Generation in Behavior-Based Safety." *Professional Safety.* 46: 27-32

the bottom are easier to influence than those near the top. No matter how motivated I am to perform any given behavior, if I am constrained by lack of proper equipment or lack of knowledge, the likelihood that I will perform the behavior is reduced. The hierarchy is useful in finding the specific source or sources of performance that need improvement. In the case of the behavior we have just analyzed, the performer has adequate equipment and facilities to spend time on safety activities. The analysis shows us that he lacks knowledge, awareness, and motivation. Knowledge is a straightforward training issue. No consequences of any kind will make a difference when the fundamental problem is lack of knowledge. The same applies to equipment and facilities.

If the issue is lack of awareness or motivation, feedback is likely to be effective, but for different reasons. If the employee is motivated but lacks awareness, feedback serves as a simple reminder, telling the person that the behavior is occurring. Let's say you are motivated to drive within the speed limit, but exceed it unknowingly. Simply getting the information that you are exceeding the speed limit will change the behavior. (Note the success of miles per hour feedback devices along roadsides as an example of this.)

On the other hand, if the actor has awareness but lacks motivation, feedback will likely have positive effects. Behaviors that occur from lack of motivation are the hardest to change, but when they are within the control of the person, they are very responsive to soon/certain/positive feedback. Behaviors that occur from lack of equipment and facilities, or knowledge, are more subject to antecedent interventions, while those related to awareness and motivation require consequences in order to be influenced effectively.

Now we can develop an action plan, Step 3. It draws from what we have learned and systematically puts antecedents and consequences in place.

Table 5-1 shows a senior leader's sample action plan.

Spending Adequate Time on Environment Health and Safety (EHS) activities		
ACTION	WHO	BY WHEN
Have an alignment meeting with direct reports and determine what safety activities are worth their attention and need their time.	Senior Leader	
Determine what we want them doing in those activities.	Senior Leader and Direct Reports	
Assign time on the calendar to complete these activities.	Managers	
Get everyone to assess their competencies in performing the identified activities. Then have another person within the organization evaluate their performance, capturing pluses and deltas each time the manager performs those activities and follow the observation with a discussion. This discussion allows me to give positive feedback on the pluses and direction on the deltas.	Managers	
Commit to doing some direct observation of things my direct reports would most like to improve, and schedule it on my calendar.	Senior Leader	

Table 5-1. Sample action plan.

Action Plan Follow-up

In 30 days, monitoring data will be reviewed for compliance with desired behavior (consequence intervention).

Example 2: Changing Behavior at the Middle Management Level

Consider a behavior more likely to be found at the supervisor/middle manager level — 'Failure to read all incident reports on a timely basis.'

> **Step 1:** Following the same model as in the previous example, we use this first step to state the desired behavior in the negative: failure to read all incident reports. Figure 5-4 shows the analysis.

A	B	C	
Not available	Failure to read all incident reports on a timely basis	Make uninformed decision	LU-
I'm traveling		Miss a prevention opportunity	S/LU-
In a hurry		Unpleasant surprise	S/LU-
No one else does		Get other work done	SC+
Trivial & Boring		Save time	SC+
Priority		Avoid boredom and irritation	SC+

Figure 5-4. Step 1, analyzing the undesired behavior.

This analysis shows that the consequences favor the undesired behavior. By failing to read all incident reports, the supervisor or manager actually saves time, is able to attend to other job duties, and avoids inconvenience. The consequences that might influence the behavior in the desired direction — further injury or negative feedback from the boss — are of the weakest kind.

Taking this example further, we move on to **Step 2:** analyzing the safe behavior. We now state the behavior in the positive: *reading all incident reports*. Now we want to list new antecedents that trigger or elicit this behavior. The analysis is shown in Figure 5-5:

A	B	C	
Leader's superior makes incident reviews a priority	Reading all incident reports on a timely basis	Leader's superior provides feedback for completing reviews in a timely fashion	SC+
Establish a standard timeline for the review process			
Install a comprehensive routing system with built-in timing requirements for review		Direct reports are recognized for innovative action plans or changes	LC+
Organization recognizes innovative or especially impactful action items coming out of reports		Reduced Injury Frequency rate	LU+
Provide training on how leaders can add value to incident investigation process			

Figure 5-5. Step 2, analyzing the desired behavior with new antecedents and consequences.

The next step is to develop the action plan, starting with new antecedents that can be added to trigger the desired behavior:

- Make incident reports available electronically
- Establish a clear hierarchy of who reviews what kind of incident
- Set aside regularly scheduled times to review incident reports

Next, we list new consequences to be delivered for the desired behavior:

- Establish timely response to action items generated by review of incident reports
- Set up monthly meetings to review what actions have been taken, and what it still outstanding
- Include incident report review and response in supervisor/manager performance reviews

Step 3: Finally, we develop an action plan (Table 5-2) to put in place the new antecedents and consequences that now favor the desired behavior.

Reviewing all incident reports on a timely basis		
ACTION	WHO	BY WHEN
Put incident reports into an electronic format.	Safety Department	
Establish a new requirement for following up on incidents: When an incident occurs, the moment it reaches the supervisor/ manager electronically, the people who are going to review the report schedule an hour on their calendar within 48 hours and include their leader on the calendar schedule even if that person is not going to review it.	Plant Manager	
Establish a hierarchy regarding who will review what kind of incident. Include a checklist on the top of each incident report that people have to check off as they review it, so the leader can see who has checked it off.	Plant Manager/ Department Managers	
Establish a monthly review of incident reports, not to go over the incidents themselves, but to ensure that everything has been read and checked off, that all actions are taken or in progress, and to see if there are any learnings that need to be shared.	Department Managers	

Table 5-2. Sample action plan.

It is interesting to note that the same patterns are found in both examples. Short-term consequences are powerful for the behaviors we would like to influence.

Putting Behavior Analysis to Work

Understanding the necessity for behavior change does not limit the scope of safety interventions; it provides a common basis for them. Any safety improvement initiative will require behavior change to assure success. The strategy is to integrate ABA principles in order to form the most efficient overall improvement strategy. Changing the culture, using cognitive methods, identifying and upgrading skill deficits of leaders, getting to the roots of motivation for safety improvement, uncovering the "emotional" component of safety, and finding ways to enhance overall safety leadership capability, are all essential elements of improving safety leadership.

Behavioral analysis is a very useful tool whenever execution is critical to success. Leaders will improve effectiveness by looking at what needs to be done in behavioral terms, using ABC analysis to develop understanding and action plans, and following up with the use of soon/certain/positive consequences.

• • • • •

Chapter 6

The Effect of Cognitive Bias on Safety Decisions

This chapter was written in collaboration with Rebecca Timmins

- Research findings on cognitive bias

- Tragedy on Mount Everest in 1996

- Applications to the organizational safety leader

- Understanding cognitive bias

- A manufacturing safety example

- Putting knowledge of cognitive bias to work

Every day, organizational leaders make decisions under a variety of pressures. They must satisfy different constituents and meet budgets and schedules, while at the same time managing the risks inherent in the Working Interface. Unfortunately, decisions made under the pressures of meeting organizational objectives often turn out to be dead wrong.

For example, often it's only after the investigation has been completed that we consider what alternative courses could have been taken, or what other decisions could have been made that would have prevented the accident. We then realize *we made judgments that were incorrect — and unnecessarily so.* Decisions were based on flawed judgments, usually concerning our assessment of future probabilities.

How likely is it that foam falling off the fuel tank will puncture the wing of the Space Shuttle? Is the use of lock-out tag-out procedures adequate to ensure workers will not contact energized equipment? Are the systems designed to control runaway chemical reactions sufficient to avoid major incidents?

Underlying each of these questions a leader must make accurate judgments about future probabilities, *judgments under uncertainty.* How are these judgments made? Do we rely on past experience to answer these kinds of questions — "we've always done it that way and nothing has happened so far" — or do we require rigorous analytic methods to demonstrate that it's safe? What kind of information does it take to cause us to respond? After the accident, it seems clear what should have been decided. But if we look carefully at what we knew *before* the accident, too often *we had all the information we needed to make a safe decision, but we didn't pay attention to it.* Why not?

Research Findings on Cognitive Bias

A rich scientific literature exists in cognitive psychology on cognitive bias that can help to answer these questions. It turns out that *human beings tend to make inaccurate judgments about future probabilities, in predictable ways.* These tendencies toward faulty judgments are called "cognitive biases," and we know quite a bit about them.

This chapter's explicit purpose is to provide a basic understanding of cognitive bias from both a practical and an application-oriented perspective. It is not intended as a source of in-depth knowledge but rather to provide a helpful framework to those who practice leadership. It aims to describe why understanding the basics of cognitive bias is helpful to leaders in the decision-making process. For more in-depth information, check the references provided.[1]

Tragedy on Mount Everest in 1996

On May 10, 1996, five mountain climbers perished in an attempt to summit Mount Everest. Two were world-renowned mountaineers Rob Hall and Scott Fischer, both skilled team leaders with vast experience climbing at high altitudes. Hall was the leader of the Adventure Consultants expedition and had successfully guided nearly forty climbers to the summit over the previous six years. Fischer, leader of the Mountain Madness expedition, had only reached the Everest summit once, but had a reputation as a skilled high-altitude climber. Climbing Everest has always been challenging and potentially dangerous: since 1922, more than 160 people have died attempting to reach the summit. There is no single reason for the 1996 tragedy; instead, there is a combination of causes, including team functioning and flawed decision-making.

The Mount Everest tragedy has become an exemplar of safety-related decision-making and the impact of leaders on shaping behavior while balancing competing pressures within their organizations. It shows how the words and actions of leaders shape and reinforce perceptions and beliefs, and lead to actions, some of which are wrong.

[1] Hammond, J. S., R.L. Keeney, and H. Raiffa. 1998. "The Hidden Traps in Decision Making." *Harvard Business Review.* 76: 47-58

Evans, B. 1990. *Bias in Human Reasoning: Causes and Consequences (Essays in Cognitive Psychology).* London: Psychology Press.

Kahneman, D., P. Slovic and A. Tversky. 1982. *Judgment Under Uncertainty: Heuristics and biases.* Cambridge: Cambridge University Press.

In 2002, California Management Review[2] published a case study on the Mount Everest tragedy, showing the role cognitive bias played in the accident. This case is a particularly good example to consider because it occurred within a complex system with multiple interactions, much like the ones organizational safety leaders face.

As background, the situation that existed prior to the Mount Everest tragedy can be characterized as follows:

- Mountain-climbing was acknowledged as a challenging and potentially dangerous sport — climbing Mount Everest was particularly dangerous;

- The expeditions' leaders, Rob Hall and Scott Fischer, respectively, were highly-skilled experts. Their reputation as guides was exceptional and they had a history of successful ascents at high altitudes;

- The clients (the participants in the climb) had invested significant resources (as much as $70,000 each) and expected to summit Everest;

- The clients had prepared for the physical rigors of the climb and were confident in their abilities;

- There were agreed-upon guidelines among both the Adventure Consultants and Mountain Madness expeditions to help manage risk — for example, a "two-o'clock rule" stated that "if you're not at the summit by 2:00pm at the latest you must turn around";

- There had been a recent history of favorable climbing weather.

Of the dozens of climbers ascending the summit that day, only six reached the top of Everest by 2:00pm. Around that time, four of the remaining climbers abandoned their bid for the summit in order to begin their descent. Hall and Fischer, however, continued on with fifteen others, with some arriving at the summit as late as 4:00pm. When unexpected bad weather developed, the

---

[2] Roberto, M.A. 2002. "Lessons from Everest: The Interaction of Cognitive Bias, Psychological Safety and System Complexity." *California Management Review*. 45:136-158

climbers were overcome by poor visibility, resulting in five fatalities among the two expeditions. The role that cognitive bias played in making fatal decisions was significant. Three types of biases were particularly noteworthy:

1. **The Overconfidence Bias:** People tend to exhibit overconfidence when they have abundant data that they believe is true, even if it has not been demonstrated to be true. Both the guides and their clients exhibited this bias: they felt confident even though the data on success in Mt. Everest ascents did not justify it.

2. **The Sunk Cost Bias:** People tend to make choices that support past decisions and escalate their commitment to a course of action they have invested in — even when there is contradictory data. The guides and clients also made this mistake: after investing so much in the climb, they did not follow the 2 o'clock rule to which they had previously agreed.

3. **The Recency Bias:** When making decisions, people tend to pay more attention to data that is recent and easy to recall. There had been a recent history of very favorable climbing weather, which overshadowed the known probability of violent storms.

Applications to The Organizational Safety Leader

For the organizational leader it's normal and usually okay to make decisions without purposeful consideration of cognitive bias. But there are critical decision points at which cognitive bias can be disastrous. So, it's important that the safety leader learn to appreciate the role cognitive bias can play.

Cognitive bias can cause leaders to underestimate exposure risk and overestimate the capability of systems to mitigate hazards. Any one small decision may be insignificant by itself, but a series of small decisions can create a path to disaster.

Understanding cognitive bias improves decision-making. Knowledge of cognitive bias positions the leader to intentionally question and look for biases that are harmful and could lead to poor decisions. Understanding cognitive biases positions the leader to give due consideration to them when making decisions — especially crucial ones.

Understanding Cognitive Bias

Cognitive bias allows us to establish shortcuts that simplify decision-making. We are prone to cognitive biases to help us accomplish two important things: 1) to simplify the complex world and make it more predictable; and 2) to help us understand new information by making it consistent with past information. These are really shortcuts in our thinking and we rely on them for various reasons: sometimes fully considering all data to reach a decision takes too long, sometimes we do not know how to find the right solution, and sometimes we lack complete data.

Cognitive bias is distortion in the way we perceive the world, but it seemingly helps us move forward when confronted with complex decisions. It is a form of mental error caused by our tendency to use simplified information-processing strategies. Biases are common to the way human beings make judgments. There are many types of cognitive bias. The following are a selection of more common ones:

- **Anchoring:** Giving disproportionate weight to the first information received — initial information anchors subsequent judgments

- **Attribution:** Associating success with personal ability and failure with bad luck or chance

- **Fact/Value Confusion:** Regarding and presenting strongly held values as facts.

- **Overconfidence effect:** Feeling overconfident in the face of an abundance of data

- **Order of effects:** Remembering data more easily at the beginning and end

- **Recency effect:** Being partial to data that is most recent and easiest to remember

- **Redundancy:** Increasing the confidence level as the data becomes more redundant

- **Rosy retrospection:** Looking back and remembering the "good times"

- **Sample bias:** Placing high value on a small sample that is flawed due to inadequate sampling technique

- **Selective perception:** Seeking data that will confirm your views

- **Status quo bias:** Preferring alternatives that support the current conditions — it's a safer strategy and involves less personal risk

- **Sunk cost effects:** Making choices that support past decisions, even when the choices appear no longer valid

- **Wishful thinking:** Preferring the decision because the outcome is desired

Each of these biases can distort decisions on their own. Even more dangerously, they can confound each other and intensify the distortion — for example, a decision is made quickly because of production pressure and then becomes established as the status quo. Or perhaps overtime decisions are made in a specific situation, but continue to be made due to the "sunk costs" bias even though inappropriate down the road.

Cognitive bias can help explain why capable people make poor decisions — why a skilled operator will "filter out" hazards (overconfidence), why a senior leader will not terminate a poor performer he or she has coached without seeing any performance improvement (sunk cost effect), or why a leader will "live" with a safety system that delivers weak safety performance (status quo).

A Manufacturing Safety Example

A manufacturing facility is planning a major expansion in six months during the slow production season for this cyclical product, at which time it will be replacing an aging production line with completely new equipment. The annual safety audit of the plant recommends that emergency shutdown interlocks be installed on the aging line because this would have prevented some serious injuries at other plants in the industry. However, due to the age and complexity of the equipment and the design and layout of the equipment and the plant, installing the new interlocks would require shutting down the line for at least a week and the line is being run around the clock to meet demand during this busy season.

The plant manager reviews the results of the safety audit with his manufacturing, maintenance, and safety managers. He explains the problems that would arise if he shuts down the line and misses production targets. The safety manager agrees with the audit recommendation but reports that this plant has never had the type of incident that the interlock would prevent. The manufacturing manager and maintenance manager agree that meeting the production target will be impossible if the line has to be shut down, and since the line will be shut down in six months anyway, recommend that the shutdown be deferred. Three weeks later a serious injury occurs that would have been avoided by an emergency shutdown.

Why didn't these managers take action, after being explicitly told that other plants had experienced serious incidents? How are they able to "live" with this situation? It is easy to say they just didn't care about safety or just didn't value their employees, but in fact, cognitive bias can "trap" people into poor decisions even when their intentions are good. In this case the managers were unduly influenced by recency bias — they had operated this line safely in the past. The plant manager had created bias among the others through anchoring — beginning the discussion by reviewing the problems that would be caused by a decision to shut down the line. They probably were influenced by both the status quo and the wishful thinking bias as well.

Putting Knowledge of Cognitive Bias to Work

How does the successful safety leader put this knowledge to work? A successful decision-maker uses his working knowledge of cognitive bias to self-monitor and to be on the alert for cognitive bias during the decision-making process. For the safety leader — where choices impact how successfully exposure to hazards is controlled — it can be critically important.

The following ideas may be useful to help leaders put knowledge of cognitive bias to work:

- Whenever you face a decision that requires the assessment of future probabilities, ask yourself what forms the basis of your assessment. Do you have data to inform your assessment?
- Gather data from various sources — seek to widen your frame of reference and look at the problem with fresh eyes
- Intentionally seek out data that both supports and discounts your theory — look at the problem from different viewpoints
- Identify a credible person to play devil's advocate
- Define alternatives clearly — the status quo is never your only option
- Consider the problem on your own first, then gather input from others
- For all decisions with history, verify that you are not giving undue consideration to sunk costs
- Study the research on cognitive bias and require your safety leaders to do the same

Perhaps most importantly, the successful leader must understand that the culture influences the extent to which cognitive bias is allowed to flourish. Does the organization sustain a value for open and free-flowing communication? In such an environment, people feel free to say what's on their minds, to speak up and challenge the assumptions that underlie biases.

How could the decision-making have been different in our example of the manufacturing plant? The plant manager, with awareness of cognitive bias, could have put the question on the table without creating anchoring. He could have asked for data on the extent of exposure at his plant versus others that had had incidents, and on alternative control methods. As the group narrowed in on a decision, he could have asked someone to argue for an alternative. Perhaps most importantly, long before this discussion ever arose he could have created a culture in which each of his subordinates was encouraged to point out possible cognitive bias "traps" whenever issues were being discussed.

Summary

Cognitive bias is a fascinating body of research with highly meaningful implications for safety leaders. Knowledge about it won't change every decision you make or cause you never to err, but it will improve your decision-making. Small gains in safety-related decision-making turn into very large gains in safety performance.

• • • • •

Section 4

Engaging Employees

Engagement is the core mechanism the effective safety leader uses to bring safety efforts to life and sustain them. But each employee level must be engaged in the most appropriate way — the senior leader requires a different approach than the front-line employee or the supervisor. In this section we look at how to engage each level.

Chapter 7

The Role of Executive Coaching in Leadership Development

This chapter was written by Jim Huggett

- Executive coaching: From remedial to developmental

- A behavioral approach to leadership

- The coaching process: Behavioral and contextual

- Step One: Understanding the context

- Step Two: Clarifying the client's unique point of view

- Step Three: Gathering the data and writing a report of findings and recommendations

- Step Four: The plan

- Step Five: Implementing the plan

- Step Six: Assessing the impact

- Coaching for safety leadership

Executive Coaching:
From Remedial to Developmental

Not long ago executive coaching was viewed as a remedial strategy; coaches were most often engaged to support executives struggling with leadership or relationship issues. Over the past decade, however, a remarkable — and positive — transformation has taken place. What was once seen as a means of "fixing" broken executives has, in many organizations, become a vehicle for developing the general leadership skills of executives and supporting them as they applied those skills to achieve specific leadership goals, including creating an organizational culture that promotes and embraces safety. Many organizations now assign a coach to all their senior leaders and, in some cases, many of their high-potential upper-level middle managers. What was once viewed as the grown-up equivalent of being sent to the principal's office has become a highly valued executive "perk" and coaching has become an important tool for developing leaders at senior levels of the organization.

But with the transition to coaching as a developmental strategy, the nature of executive coaching has had to adapt to meet the needs of executives seeking to enhance their overall effectiveness as leaders. With remedial coaching, the issues, or at least the symptoms, are generally apparent, and the coach is usually engaged to help the leader address a specific behavior or behaviors that are causing difficulties. With developmental coaching, the coach becomes the "voice of the organization," helping the leader to understand how his or her portfolio of behaviors impacts key stakeholders and either furthers or impedes the ability of the organization to meet its business goals. While this may include modifying less constructive behaviors, it also frequently includes developing strategies to leverage the leader's strengths to greater advantage.

A Behavioral Approach to Leadership

One of the reasons leadership often seems so mysterious is that it is frequently discussed at the "characteristic" level; leaders are "charismatic," "compelling," "visionary," or even "Machiavellian." While describing leadership in characteristic terms may be generally informative, it is of little value to individuals trying to figure out what they can do to improve as leaders. Telling someone he needs to be more "charismatic" does little to help him figure out what he needs to change to improve his day-to-day leadership.

However, all we ever know about others is based upon what they do or say — their behaviors. A leadership characteristic, then, is a perception we come to through direct or indirect observation of a leader's behaviors. If we can break down the esoteric characteristics into the underlying behaviors, we can then begin to help leaders think about how to change behaviors to enhance their overall effectiveness.

A recent experience we had with a senior-level leader illuminates this point. In confidential interviews on behalf of the leader, reports frequently described her as "indecisive," a characteristic. As we probed further, we learned that employees in meetings in which decisions were anticipated, observed that the executive would listen quietly and with minimum engagement (behaviors). When the meeting time expired, she would simply thank everyone and end the meeting (behaviors). The attendees, having anticipated that a decision would be forthcoming in the meeting, interpreted the leader's behaviors as "indecisive".

Just informing the leader that she was perceived as "indecisive" would not have been constructive. In fact, it could have been counterproductive if she responded by making ill-informed decisions just to address the perceptions of others. But once the underlying behaviors that led to the perception were understood, only simple behavioral adjustments were required to change the way others viewed her.

After reviewing the behavioral interview data, we developed a plan to address the perception of indecisiveness. At the beginning of any meeting in which the possible outcome was to discuss or make a decision, she opened the meeting by identifying and articulating the decision to be made, how it would be made (i.e. consensus, majority vote, leader with input, leader without input) and what information was required to make the decision. Most importantly, she clearly articulated when the decision would be made. In other words, rather than trying to become more "decisive," she became more *definitive* in communicating her decision-making process. The impact was almost immediate; follow-up interviews with her reports a month later indicated that by simply becoming more transparent in decision-making, she was able to change the perception of herself from "indecisive" to "very decisive."

Can people really change? Can working with a coach help an executive change behaviors developed and ingrained over years? The answer is yes, of course. People change their behaviors all of the time. If you drive to work the same way for thirty years and a faster route becomes available, it's easy to change your thirty-year routine. It is more difficult to change the underlying value you hold that led you to evaluate your behavior and decide to change it. You didn't change your value of wanting to get to work as quickly as possible, you changed your behavior because you found a new behavior that provided an outcome that better supported your value.

The Coaching Process: Behavioral and Contextual

As the coaching process has evolved to become an important tool for executive development, another change has taken place. Leaders have begun to require that the coaching process be tightly linked to the business goals of the organization. In the past, many coaches saw their role exclusively as providing a "sympathetic ear and a little advice" to their clients. Without minimizing the value of that sympathetic resource, developmental coaching, when done well, is now highly contextual. It requires that both the leader and the coach have a firm grasp of how the executive's leadership behaviors support or impede his or her ability to drive the organization's agenda. The most successful coaching relationships are now based upon a well-structured and data-driven approach. We measure a coaching relationship's success by whether it helps to ensure the organization achieves its business goals.

Step One: Understanding the Context

In most developmental coaching relationships, the overarching goal is to help the leader understand how his behaviors impact reports, peers, and managers, and to influence his ability to meet personal and organizational goals. Logically, one of the first things the coach must understand is the context in which the leader must lead. Leadership is, of course, both highly situational and highly contextual.

The most common example of situational leadership is referred to as the "foxhole" dilemma. In most situations, gathering input from others before making a decision would normally be viewed as a positive leadership behavior. However, when the bullets are flying — literally or figuratively — the time-consuming process of asking others for input may, at the risk of understating the obvious, be inappropriate. Simply stated, successful leadership will probably look different in a foxhole than it does in a boardroom.

The organizational context in which the leader is leading is equally important. What is acceptable in one organization may be inappropriate or ineffective in another. For example, in our experience, as leaders become more senior, their focus should shift from day-to-day tactical activity to a more strategic perspective. However, we recently worked with an organization characterized by large, complex high-risk/low-margin projects. Even at the most senior levels, executives were expected to have detailed, hands-on knowledge of each project. Responding to a detailed question from the CEO with "John handles those details — I'll check with him and get back to you" could be a "career-limiting" event. Had we coached our clients to fit our own bias, we could have easily coached them out of alignment with that organization.

Step Two: Clarifying the Client's Unique Point of View

The coaching process generally starts with a meeting between the coach and the leader. The initial meeting serves two purposes. The first is to explore the goals and objectives the leader must achieve to meet the requirements of his role. These are usually (although not always) well documented and measured. But in addition, the coach should be looking for the degree to which the leader's *personal* goals and values fit with, and support, his professional goals. While this issue is often overlooked in coaching relationships, it is a critical factor in the leader's ability to succeed in his role.

We worked with a CFO recently who was required to provide support to a number of widely distributed operating units. His job, by definition, required him to spend at least 60% of his work week on the road. While his personal values and objectives certainly included success in his career, they also included spending substantial time at home sharing responsibility for raising his two preschool children.

The conflict between the two roles was a constant source of stress in his life and, ultimately, led to a sense that he was performing neither role very well. As he began to recognize the conflict, he was able to make decisions to better align the roles and minimize it. In this case, he was able to petition his organization for a role with less travel. He traded his fast-track career objective for a role that provided better balance and less stress.

Step Three: Gathering the Data and
Writing a Report of Findings and Recommendations

Once coach and leader are clear regarding both personal and professional goals, we typically recommend conducting a series of interviews. Since we are looking at the leader's overall impact within the organization, a 360 assessment usually provides the most comprehensive picture. We normally conduct confidential interviews with the leader's boss and a number of direct reports. Also, we frequently recommend that we interview several of the leader's peers; as leaders become more senior, their ability to work well with, and influence others, across organizational boundaries often becomes at least as important as their ability to work with their reports.

The challenge in the interview process is to get beyond "characteristics" to ensure that the data gathered is truly behavioral. When asked, "What comes to mind when you think of Mr. Jones in his leadership roles"? most respondents will naturally respond with a characteristic — he "really cares about his people" or "she's arrogant." But the leader needs to know what behaviors led respondents to perceive that she is "arrogant" (for example, she doesn't listen when others speak, or doesn't make eye contact). If the Findings and Recommendations report doesn't provide this kind of information, it will probably make him feel unhappy — or pleased — but it will be of limited value in helping him improve.

In addition to behaviors, another element of the data gathering is equally important. We all have a large portfolio of behaviors, only some of which impact our ability to lead. When gathering information regarding characteristics and behaviors, it is also important for the coach to determine the impact those behaviors have on others.

We worked with an executive who was observed to become short-tempered on Friday afternoons as he tried to tie up loose ends before the weekend. His reports made a conscious effort to avoid him during that time. But further probing around "impact" found this general consensus: "We avoid him on Friday afternoons but it's no big deal. If we need something, we wait until Monday." Hence, while the impact of this behavior was not what the leader desired, when we developed a coaching action plan, addressing this behavior fell to a lower priority due to its limited impact on his organization.

With a clear picture of how the leader's behaviors impact those he must lead and how those behaviors influence the ability of the organization to meet its goals, the coach can begin to consider recommendations. But, rather than suggest the leader should change this or stop that, an effective coach will review with the leader the consequences that result from a specific behavior or pattern of behaviors. The recommendation should suggest where to look to assess impact rather than "what you should change."

Recently, we worked with a new CEO who tended to become caustic with his reports in meetings when he felt they were acting or speaking in ways that promoted the interests of their own departments over the good of the organization. According to the leader, and confirmed by our own observations, this parochialism was a significant — and destructive — characteristic of his senior team. His behavior was an intentional strategy aimed at reducing the self-interest and sub-optimization that characterized his organization.

As we gathered data from the senior team on behalf of our client, we learned that his reports were aware of his position on parochial attitudes and soon learned to avoid comments that could be interpreted as self-serving. Therefore, his behavior was having the effect he intended. However, we also learned that his potential for caustic responses led many members of his team to avoid challenging his position on any issue. Thus, the leader was getting what he wanted, but he was also getting an unintended consequence; he was shutting down the input and participation of the members of his team.

Our recommendation was not that he should stop what he was doing. Rather, we suggested a review of the assessment data to fully understand the impact of his behavior. We then discussed alternative behaviors that would have the desired impact without the unintended, and undesirable, consequences.

It's important to note that this approach is nonjudgmental. The coach is not embedding his or her values into the process by suggesting that "this behavior is good" or "that behavior is bad." The underlying question is whether, on balance, a behavior supports the goals and values of the client and the organization or gets in the way.

Step Four: The Plan

With a clear perspective of the leader's personal and professional goals and an understanding of how his or her behaviors either support or impede the ability of the organization to achieve those goals, it is time to develop a plan to close any gaps and leverage the strengths.

There are a several important points here and the first is the most obvious — this is the leader's plan, not the coach's. The coach can offer suggestions and help to document the plan. The coach can also play an important role in supporting the implementation and evaluating the impact. However, it is up to the leader to take ownership of the changes and to drive the plan to completion.

Second, the plan should address no more than the top three or four issues, prioritized by the impact they have on the organization. Other issues the client would like to address, but with less impact, should be queued up for future efforts. Behavior change is hard work. A plan that is overly ambitious and tries to do too much can easily become overwhelming and may never be completed.

Where appropriate, the plan should also point out opportunities for the leader to take advantage of strengths identified in the assessment. For example, if the assessment identifies a lack of clarity around strategy but strong large group communication skills, using the latter to enhance the former plays to the leader's strengths.

The plan doesn't have to be complicated — in fact, the simpler, the better. It should include a concise description of the action steps (behaviors) the leader will employ, the issue or gap the actions are intended to address, who will do what and when, and how the coach and leader will measure the impact.

Step Five: Implementing the Plan
(Finally, the Real Coaching Begins)

At this point, what actually transpires in the coaching relationship depends on the gaps identified in the assessment and the subsequent planning process. The coach's role is to support, suggest, measure, cajole, nag, and provide input. It is the leader's job to "do" — to make the changes that will ensure the objectives established in the plan are met.

Some possible roles for the coach include observing the leader in situations in which he or she will be applying the new behaviors and providing both corrective suggestions and positive feedback. The coach can also help the leader to think through methodologies, techniques, meeting agendas, and communication tools that will help him achieve the desired outcomes. And, of course, provide a "sympathetic ear and a little advice" when called for.

Step Six: Assessing the Impact

There is no magic in planning. Most of the time, a thoughtful plan, well executed, will get the desired result — but not always. Once the coaching plan has been implemented and sufficient time has elapsed to allow it to impact the stakeholders, the coach should "circle back" to see if the leader is having the desired impact. This second round of data gathering provides two benefits.

First, it informs additional steps or actions the coach and leader can take to continue to improve the leader's impact. With the additional data, the leader can make further refinements or move on to lower priority issues.

Perhaps more importantly, because the coaching process includes a specific re-evaluation step, it provides a "consequence" for the leader's efforts. Simply knowing the coach is going to circle back and ask the initial respondents "How is he doing?" creates incentive for the leader to stay focused and to avoid reverting to old behaviors during the implementation phase.

Coaching for Safety Leadership

Creating a truly safe working environment requires that two separate but related elements come together in support of safety. The first is the most obvious: does the organization have the right safety enabling systems (practices, policies and procedures) in place aided by the organization's supporting systems (reward and recognition plans, compensation plans, communication strategies, etc.)? Many organizations emphasize the enabling systems side of safety and are quite good at it. Also, over the past decade recognition of misaligned performance drivers within organizational supporting systems has become more prevalent and organizations have taken steps to eliminate performance drivers that encourage unsafe behaviors. An example of one such driver would be piecework compensation plans in dangerous operating environments.

The second element in the creation of a safe work environment is less visible but perhaps more important: the organization's culture. Does the organization have a culture that demonstrates a profound intolerance for exposure to hazards and truly believes that safety is everyone's responsibility? Getting the right safety systems in place is good management. Creating a culture in which exposure to hazards is not tolerated — where workers and supervisors work together to keep the Working Interface safe — is good leadership.

Some work we did recently with the general manager of a large US-based oil refinery illustrates the importance of this point. Over the past few years, the GM had aggressively addressed many of the safety enabling systems in the organization. He brought in an experienced Health Safety and Environmental (HSE) executive at a senior level and gave him the resources to evaluate and improve the organization's safety enabling systems. But after several years of effort, OSHA's recorded injury rates showed no consistent and significant downward trend.

In frustration, the general manager asked us to take a look at his safety program and make suggestions to help him "re-energize" his efforts. We initiated our analysis with a safety audit of his systems and procedures. The data indicated that the organization had a solid foundation of safety enabling systems in place. His new HSE executive had done an excellent job.

We then used the Organizational Culture Diagnostic Instrument (Chapter 4) to look at the underlying attitudes and perceptions within the organiza-tion related to safety. Our findings here were more illuminating. The data indicated that there seemed to be a somewhat fatalistic attitude in the plant regarding safety. Many employees, supervisors, and managers seemed to feel that it was "risky work" and that "accidents [were] just going to happen." Instead of a total intolerance for exposure to hazards, there seemed to be almost a cultural acceptance that risk was part of the job. In that kind of an environment, safety shortcuts almost seemed to be encouraged.

There is no doubt that the general manager is responsible for the culture of his organization, and we have discussed the leader's role in creating a safety culture in other parts of this book. The relevant question in a discussion of the role of executive coaching is what impact the leader's personal behavior has on culture and how coaching can help him or her sort that out and develop a portfolio of behaviors that promote his safety agenda. With this in mind, the GM asked for an executive coach.

After getting clear on what the general manager was trying to accomplish, the coach interviewed a series of managers, supervisors, and employees throughout the organization to see if he could pinpoint behaviors that either supported and enhanced the GM's safety goals or were counterproductive. The data collected showed some interesting perceptions across his organization.

The GM had been in his role for three years. Generally, the data indicated that while some of his behaviors communicated a lack of resolve regarding safety issues, the real culprit was a lack of behavior. The current safety culture was long-standing, and the GM's behaviors had done little to influence it. So, it wasn't so much that his behaviors were communicating the wrong message but that they were insufficiently strong and positive to change the way people perceived safety within the refinery.

We broke down the leader's portfolio of behaviors into positive and negative. Positive behaviors, that is, those that were perceived to support his safety goals, included:

- Hiring the new and aggressive EHS manager and financially supporting his efforts to upgrade the organization's safety systems

- When addressing employees in presentations, newsletters, etc. he usually took an opportunity to express his commitment to safety

- His direct reports all had a safety component in their annual goals

Behaviors that were seen as inconsistent with a commitment to safety included:

- Allowing latent exposures to hazards to be inconsistently addressed if they did not lead to a "recordable" incident. People perceived a lack of consistency and diligence in how leaders and the GM responded to exposures to hazards

- While safety was often a topic in his presentations, more often productivity was emphasized, leaving most people with the perception that safety was important as long as it did not inhibit production

- Managers who were known to have only a mediocre commitment to safety were promoted

- The dialogue regarding safety tended to fall off after a period when no accidents had occurred

- The GM had been observed touring a "bunch of suits" through a hardhat area without requiring them to wear hardhats. (It is amazing the number of times we heard this story during the interviews.)

The data indicated that he was doing some of the right things — and not too many of the wrong ones. On balance, however, his behavior did not communicate sufficient commitment and intensity to challenge the old culture. And while it is unrealistic to believe his behavior alone could change the culture, without his leadership, nothing else he or his team could do would have the desired impact. With this data in hand, the next step was to develop a plan to change perceptions regarding his commitment to safety, the critical first step to changing the climate, and eventually the culture, of safety.

Discussions with the GM revealed that he hadn't thought seriously enough about what it meant to be serious about safety, and how critical his role was to the organization's success. When he realized he was leading the organization to be mediocre in safety, he was forced to question his own values and the degree of his emotional commitment. He was quick to arrive at answers: he did value safety and was willing to take on the task of changing his own behavior in order to show that value.

The objective of the GM's behavioral action plan was straightforward. Through his own behaviors, he would communicate and promote an absolute intolerance for exposure to hazards within his organization. He sought to create an environment in which everyone knew that the organization's espoused value for "our people" was unequivocally expressed by a total commitment to safe practices. Safety shortcuts were unacceptable, regardless of the impact they had on production.

The components of the coaching plan were equally straightforward. He started by taking his leadership team offsite to develop a vision for safety. Such a vision means more than espousing a goal of zero injuries —that's just the measurement. The key element of a vision is identifying the behaviors that will be necessary from every manager and employee to create a culture that embraces safety. It responds to the question, "If you were absolutely successful in creating a culture that supports safety (and you couldn't look at the statistics), how would you know? What would be different in the way people think and act in your organization?"

With the vision as the cornerstone for every communication, the GM initiated every speech, presentation, or broadcast email by reinforcing the leadership team's commitment to safety. He increased his level of safety communication and encouraged his managers to do the same.

Next, he built a safety component into every performance plan in the organization and weighted the category above other objectives such as productivity, output, or cost management. Safety was no longer "one among many"; it was now "how we do all of those other things." In addition to creating a positive consequence for aggressively supporting the safety agenda, the effort unequivocally communicated the GM's commitment to achieving the organization's safety vision.

The GM then installed an employee-driven safety process to teach and encourage employees to observe exposures to hazards and provide feedback on reducing them. Managers and supervisors were trained by going through a similar process to the GM's to help them understand their relationship to keeping the Working Interface safe.

And, of course, the leader took every opportunity to demonstrate his own commitment — no more tours for the "suits" without hardhats.

Leadership is, at its core, the ability to get others to behave in ways that further the goals of the organization. With the support of his coach, the GM was able to identify a set of behaviors that communicated and reinforced his tangible and deeply felt commitment to the safety of everyone within his sphere of responsibility. But perhaps of more impact was his focus on his "behavioral leverage" — those behaviors that frame, inform, and reinforce the behaviors of everyone in his organization.

So, is it nature or nurture? Can a coach take a mediocre leader and turn him into a Winston Churchill? Probably not. But, like most things in life, leadership lives within a bell curve. Some leaders — only a small number perhaps — seem to be born to it and populate the right side of the bell without effort. For the rest of us, understanding how our behaviors impact others and then adjusting those behaviors to better align with our goals can certainly help us to migrate to the right and improve our overall ability to lead others. We can get better.

· · · · ·

Chapter 8

The Role of the Supervisor
in Leading with Safety

This chapter was written in collaboration
with Kim Sloat, Ph.D.

- The pivotal role of the first-line supervisor

- Communication skills — the foundation

- The power of strong working relationships

- Fair decision-making and its effects

- Alignment: Incorporating organizational values and priorities
  into day-to-day activities

- Safety contacts: Getting an accurate picture of performance

The Pivotal Role of the First-Line Supervisor

In most organizations the first-line supervisor is a key influence on organizational effectiveness. The supervisor links management and the workforce. Supervisors often have longer tenure than managers and frequently have supervised the same workgroup for many years. They often have more credibility with workers than managers. Workers look to the supervisor to interpret organizational priorities and changes and they may ask the supervisor directly about the meaning of some particular management action. More typically, the supervisor interprets management actions in light of his or her experience and passes them on in the form of directions, informal comments, and reactions to workers' actions. In many respects workers take the words and deeds of the supervisor to represent "the company." That is, worker perceptions about the organization are filtered through the supervisor.

Supervisors affect safety outcomes in several ways. First, they affect exposure to hazards depending on how well they use safety enabling systems and tools such as safety meetings, incident investigations, inspections, audits, and the identification and mitigation of hazards.

The supervisor plays a key role in addressing the range of exposures that exist in the workgroup. It is helpful to think of the degree of control that the worker has to perform work safely. At one end of the spectrum are "enabled" situations — the worker has the necessary skills, knowledge, and resources to work safely. At the other end, are non-enabled situations — the worker is unable to do the work safely. In between these extremes lie situations of varying degrees of difficulty — the work can be done safely, but doing so takes extra time and effort. The supervisor addresses different situations by using different combinations of the tools available to him.

The second way supervisors can affect safety is by communicating organizational priorities and values. In part, he does this through explicit statements. More powerfully, he communicates priorities and values by what he says and does, or does not say or do, at critical moments. For instance, suppose a supervisor is under pressure to get a key piece of equipment into operation. If he insists that safe procedures be followed even though it delays the availability of the equipment, the priority of safety will be communicated much more powerfully than by saying "Safety is number one" in every safety meeting.

Third, supervisors affect safety because their interactions with individual team members and the group as a whole affect the overall tone or climate of the workgroup. If team members consider the supervisor to be fair and supportive, they are more likely to extend themselves with respect to objectives important to the supervisor. If safety is important to the supervisor, and he has created a positive climate, team members will accept more responsibility for safety. A workgroup with a fair and supportive supervisor functions more effectively and team members have more positive relationships. In such a group, workers are more likely to look out for one another.

Communication Skills—The Foundation

Good communication skills are fundamental to a supervisor's effectiveness. Surprisingly, good communication is hard work. There are two sides to communication — sending and receiving. The messages we send and receive are filtered through our perceptions of the world, as well as our needs and feelings about the situation. That means that what we actually send — and the other receives — is not always what we intended to communicate. That we can think much faster than we can speak, listen, read, or write, adds to the difficulties. We often jump ahead and assume we know what someone intends to communicate, or our attention wanders and we miss a critical piece of information.

A supervisor can improve workgroup effectiveness by paying attention to the accuracy of the information he or she sends and receives. That can be done by checking for understanding — verifying that workers understood what the supervisor intended to communicate, and that the supervisor understood correctly what the worker meant. Effective communication is thus an active process, and takes some work. Ideally, all members of the group would verify that they understand and are understood. But not all workers are likely to play such an active role. The supervisor then has to work both ends of the communication, making sure the worker understood him and that he understood what the worker intended.

In a busy workplace, taking additional time to verify the accuracy of communication can seem like a luxury. The alternative, however, is that from time to time there will be misunderstandings, with potentially catastrophic consequences.

To ensure he was understood, the supervisor needs someone in the workgroup to say or do something to let him know he was understood correctly. For example, suppose the supervisor is supposed to go over a new safety procedure. A common approach would be to read, show, and/or pass out information about the new procedure and to ask if there are any questions. In the absence of questions, the supervisor might assume that workers understood the procedure. A more effective approach might be to ask the workgroup how the procedure would apply in specific cases. With this or some similar approach, the supervisor would be in a better position to assess understanding.

When he is on the receiving end, a supervisor can improve communication by using "active" listening skills. When listening actively, a supervisor concentrates on the key points the other person is making. Instead of assuming that he is getting it, the supervisor actively checks by paraphrasing what he has heard. The other person can then verify that the supervisor heard correctly, or clear up a misunderstanding.

Regular use of active listening has two powerful outcomes. The first, which is obvious, is a much lower likelihood of misunderstandings. Misunderstandings lead to errors, inefficiencies, and strained relationships. The second outcome is less obvious. When a supervisor uses active listening skills, the worker takes that as a sign of respect — the supervisor cared enough to make sure he understood the worker's point. Whether the supervisor agrees with the point or not, use of active listening acknowledges the worker's status within the group as a person to whom it's worth listening. A supervisor builds up tremendous goodwill with team members through effective communication skills. In turn, team members are more likely to treat the supervisor with respect and dignity. These experiences with active listening contribute to the development of effective working relationships, or Leader-Member Exchange, discussed in Chapter 4.

The quality of supervisors' communication skills varies, as does the consistency with which those skills are used. Those with less developed skills will benefit from training to improve them. For any supervisor, it can be helpful to get brief periodic feedback (from team members, peers, and the boss) on how well and how consistently specific communication skills are used. One useful approach is to ask for feedback through a 360 diagnostic instrument (see Chapter 3).

The Power of Strong Working Relationships

A long-standing definition of "management" is that it is getting work done through others. Although some supervisors also do some of the work themselves, at its heart supervision is about motivating, coordinating, and directing the efforts of other people in accomplishing organizational objectives. The ongoing challenge for supervisors is how to do this.

Relationships are a crucial aspect of leadership. A supervisor who is technically brilliant but who has poor relationships with his or her team members will be less productive than a less brilliant colleague who has excellent working relationships with team members. Understandably, the quality of supervisor-member relationships varies within a workgroup; the relationship is stronger with some members, weaker with others. However, the average strength of relationships with team members is different for different supervisors, and those with the strongest relationships will be more effective over time.

Relationships develop as a sort of exchange. To start, the supervisor gives assignments to a new member of the group. If performance is good, the supervisor gradually gives the worker more latitude. Over time, mutual trust and respect develop. There are advantages to a strong working relationship for both the worker and the supervisor (Table 8-1):

TO THE WORKER	TO THE SUPERVISOR AND THE ORGANIZATION
Support from supervisor	Worker satisfaction with supervisor
Satisfaction with work	Worker commitment to the organization
Access to resources	Overall performance by worker
Greater autonomy	Less need for close supervision
Open communication	Open communication
Influence on decision-making	Organizational citizenship behavior
Better information	Reduced turnover

Table 8-1. The advantages of a strong working relationship.

Strong relationships are based on several key elements. The principal one is trust. Trust is particularly important for team members, as they often feel relatively powerless, and uncertain about their status. They are highly alert to incidents that may have meaning about their place in the organization. They spend more time thinking about interactions with the supervisor than the supervisor does about interactions with the team member. Subordinates often read unfavorable motives or meaning into incidents that reflect no ill intentions on the part of the supervisor. If the relationship is a trusting one, it is more likely the team member will give the benefit of the doubt to the supervisor. Supervisors can improve trust by acting in a trustworthy manner, which includes:

- **Consistency:** Acting in a predictable way across time and situations

- **Integrity:** Telling the truth and keeping promises

- **Communication:** Exchanging ideas and accurate information, and explanations for decisions

- **Good intentions:** Considering team members' needs and interests, acting in a way that protects employees' interests, not exploiting others

- **Delegation and input:** Sharing control and participation in decision-making

Fair Decision-Making and Its Effects

Supervisors make decisions daily that affect the work lives of their team members. These decisions may include job assignments, work schedules, breaks and their timing, vacation, overtime, discipline, rewards, and training. When workers are represented by a union, the decision-making process for many decisions is spelled out in the contract, and the supervisor has relatively less latitude. Nonetheless, the supervisor still makes many decisions that a worker might see as favorable or unfavorable.

Outcome fairness: There are three main aspects of supervisor decision-making that a worker might focus on. The first is whether the worker gets what he or she thinks is a fair outcome. Workers compare the decision outcome with what they hoped they would get, and with what others got. For instance, a worker compares his job assignment with what he hoped for, and with what assignments others got. Inevitably, the supervisor will make decisions that make some workers unhappy. (To paraphrase Abraham Lincoln, you can please all of the people some of the time, some of the people all of the time, but you cannot please all of the people all of the time.) Over time, it would be expected that all team members would be unhappy about one decision or another. Because many people nurse grudges, it would not be surprising if the result were a workgroup of unhappy people who do not get along with the supervisor or each other.

Procedural fairness: Although some workgroups do become dysfunctional, this is not an inescapable result of the supervisor making decisions that displease some team members. That's because there are two other aspects of decision-making that affect how members view the outcome. The second aspect involves judgments about the decision-making *process*, known as Procedural Justice. (See Chapter 4 for methods of measuring Procedural Justice.) If an outcome seems unjust, but the supervisor made the decision using fair procedures, the worker is much less likely to be unhappy. The worker seems to realize that everybody will not be happy with every decision, but if the supervisor is making decisions in a fair way, over time people will get their fair share of desirable outcomes. Employees take fair decision-making procedures as a sign of respect and consideration for them. This leads to positive reciprocation on their part.

What makes a decision-making process fair? It generally has the following elements:

- **"Voice":** Making sure the concerns, values, and outlook of those affected are given consideration

- **Consistency:** Applying the same standards across people, situations, and time (at least in the short run)

- **Bias suppression:** Preventing personal self-interest or feelings about the individual from affecting the decision

- **Accuracy:** Taking into account all relevant information, ensuring it is accurate, current, and complete

- **Correctability:** Giving workers the opportunity to appeal decisions or procedures without retribution

Interaction fairness: The third key aspect of decision-making is how the supervisor treats people in the course of the process. As with fair procedures, fair treatment can help overcome disappointment over outcomes. Unfortunately, supervisors sometimes become defensive about unfavorable decisions, and the defensiveness makes the situation worse. When a supervisor treats people fairly, it is much more difficult for a worker to blame the supervisor.

Fair treatment includes relaying timely and accurate communication, listening actively, and treating workers with dignity and respect. It may also include:

- **Excuses:** Describing circumstances beyond the supervisor's control that affected the decision

- **Justifications:** Outlining the thinking behind the decision that makes it fair overall

- **Apologies:** Admitting to responsibility for the unfavorable outcome

Can supervisors improve the fairness of their decision-making? Yes. It has been shown that training targeted specifically at fair decision-making improves perceptions of fair treatment in the workgroup and increases "extra-role", or "Organizational Citizenship", behavior — workers going above and beyond the call of duty.

Alignment: Incorporating Organizational Values and Priorities into Day-to-Day Activities

Most organizations have multiple competing priorities at all times. There are relentless cost- and production-efficiency pressures. There are fewer people, and more things to do. One program after another is introduced; all are important. Time is at an absolute premium, but meetings and paperwork continue to be added to the schedule. How can these various demands, especially in safety, be met with increasingly limited time available to supervisors?

One effective strategy is to make sure that time devoted to particular activities is maximally effective, and to leverage the time to accomplish multiple objectives. In most organizations, supervisors have specific activities they are expected to do for safety. Examples include safety meetings, five-minute safety talks, pre-shift meetings, inspections, and so on. These activities were introduced at some point in the organization's history for specific reasons — someone identified a gap in the safety program, and an activity was put in place to fill that gap. Over time, the purpose of the activities commonly gets lost, and the activities start to exist for their own sake — safety meetings are held because they have "always" been held, but few people think about the organizational purpose of the meetings.

There are two ways to improve the effectiveness of time already devoted to safety. First, be clear on the purpose of the activity from a safety standpoint. Suppose we asked a group of supervisors in the same organization to write down the purpose of a given activity, say incident investigation. The items on their lists would overlap somewhat. Some items might even appear on every list, but the lists would not be identical. Even for the items that overlap, some descriptions of them would be clear and precise, and others would be vague. The result is that the various supervisors' lists of the purposes of incident investigations would be different. The various parts of the safety program will be more consistently effective if the involved supervisors and managers agree on the purpose of the parts. With a clear purpose, it is then possible to measure how well the purposes are accomplished.

The second way of improving effectiveness is to identify other organizational objectives that can be met through the safety activity. For example, leaner staffing may lead to an organizational objective to increase individual initiative. The value of safety activities to the organization would be enhanced if they also helped increase individual initiative. A supervisor might think of ways in which incident investigations might increase individual initiative. For every step in the investigation process — reporting incidents, caring for injured employees, investigating the mishap, following up on action steps, and sharing and applying the investigation results — there are likely to be opportunities for increased individual initiative. These opportunities would be not only for the injured party, but likely also for co-workers, the supervisor, and personnel from other departments.

Supervisors can enhance safety not only by conducting safety activities in particular ways, but perhaps more powerfully through the informal contacts they have with team members. In pre-shift meetings the supervisor can not only talk about the work to be done, but also have team members talk through the potential safety hazards and discuss what tools, equipment, and planning might be necessary for safe work.

Safety Contacts:
Getting an Accurate Picture of Performance

It is easy for a supervisor to overlook the most important exposures in the workgroup. Supervisors are extraordinarily busy, and paperwork and meetings consume time that could be spent in the work area. When he or she does pass through the work area, only the more obvious exposures are likely to be noticed, such as failure to wear eye protection in production areas. When workers raise issues, they are generally facility or equipment items. Such items are often important and should be addressed appropriately. Exposure to injury involves the worker in interaction with the immediate work environment. Supervisors can improve safety by concentrating on the exposures that contribute most often to injury and those with the greatest potential for severe injury. Sometimes these exposures can be addressed by changes in facilities and equipment, design, or procedures.

Engaging workers in identifying the root causes of the exposure can be helpful in understanding and addressing the root causes. This works best when the supervisor takes time to observe the safety-related aspects of work being performed and gives positive feedback for jobs done safely. Discussing exposures with the worker to determine what needs to be done to make the task safer is helpful before determining an action plan for addressing the hazard.

This approach is easiest when the organization is using an employee-driven process that focuses on improving the Working Interface. (See Chapter 9.) Information from data gathering helps to focus improvement efforts. If such a process is not in place, the need for supervisor action is higher. Even when data gathering is in use, direct interaction between the supervisor and team members about reducing exposure is necessary.

The first step is to identify the exposures on which attention should be focused. With an existing process for data gathering, data will point to the highest risk or highest exposure situations. Otherwise, various approaches can be used, including review of injury or incident reports, reviews by safety professionals, and discussions with the workgroup. Often the most important exposures will not be ones that are typically noticed in casual observation.

The second step is to develop a plan for making safety contacts. The goal is for the supervisor to have regular contact with each member of the work group while at the same time spreading the contacts across tasks and times so as to get a representative picture of the level of safety in the workgroup. A contact with each team member between once a week and once a month is a useful target. It is very helpful to keep track of who was contacted and the work observed to ensure a representative sample of the work.

The third step is to make the contacts. The objective is to focus on the safety aspects of the work being done, concentrating on the situations of highest exposure or highest risk in the work group. The contacts should be open and obvious, and discussed with the group in advance. The goal is to develop a collaborative approach to improving safety. Depending on the task, five to ten minutes of observation usually will be enough to get a fair picture of what is going well and what needs improvement. The most important part of safety contacts follows the observation — feedback and discussion. The most effective strategy is for the supervisor to start with positive feedback on things done safely. Following that, the supervisor encourages discussion on anything that exposed the worker to hazards.

The discussion of exposures is the biggest opportunity for improvement. The goal is to discover the sources of exposure and how to address them.

Finally, follow-up enhances the power of the contact. The plan developed by the supervisor and worker calls for either or both of them to take certain actions. It is critical that the supervisor does two things. The first is to complete his actions and communicate back that they were done. Follow-through may be critical to reducing the exposure and it also increases his reputation for integrity. Failure to complete agreed-upon actions will damage perceptions of integrity. Second, the supervisor must follow up to see if the team member has completed his or her commitments. If so, the supervisor can provide positive feedback, and the interactions serve to strengthen the working relationship.

Regular safety contacts send a strong message about the supervisor's value for safety. There is frequently confusion on the part of team members about the importance of safety relative to other organizational demands such as production. On the one hand, they may hear messages during safety meetings that stress the importance of safety. On the other, day-to-day events may suggest that production is the most important thing. Such confusion is more likely if safety messages are mostly given during specific times or events (safety meetings, tailgate meetings). The supervisor can reduce confusion by delivering a consistent message, and regular safety contacts are a powerful way of doing so.

• • • • •

Chapter 9

A Systematic Process for Reducing Exposure to Hazards

What the safety improvement process looks like at the worker level

This chapter was written in collaboration with
John Hidley, M.D.

- Engagement and cooperation

- Getting engaged in safety

- The safety improvement mechanism
    - Identify exposures
    - Gather leading indicator data
    - Use data to provide feedback
    - Reduce exposures
    - Renew the process

- Implementing the process

- Roles at every level

- Leadership and reduction of exposure to hazards

- Best Practices

- Getting started

- What to expect

Systematically reducing exposure to hazards takes the active collaboration of leaders and workers. Workers have a role to play in understanding and supporting the process and they benefit most directly from its benefits. Leaders play a different role: creating a culture in which safety is valued. Both are essential for optimal results. To reduce hazards most effectively means using a methodology to identify them, and then removing or controlling them. This is best accomplished when employees at all levels are actively engaged in getting safety right.

Engagement and Cooperation

Getting safety right requires everyone's *willing* participation — and more than their participation, their active and wholehearted engagement. Whether such engagement comes easily or is almost impossible to achieve depends on the atmosphere leadership creates. The key to safety success is for leaders at all levels to cooperate to achieve the common goal of making the workplace safe.

We reject the dichotomy, sometimes assumed in discussions about safety, that the safety interests of labor and management are different and opposed. Safety improvement is a process built on explicit cooperation between workers and leaders of all kinds. Everyone has a role to play and the robustness of the safety improvement process is in direct proportion to how well everyone does his or her part.

Workers are usually present at the point of exposure and know why things are done the way they are. Their involvement may be critical to identifying what needs to be done differently. And because they perform the work, their genuine support of changes is essential to successfully sustain improvements. But workers cannot make decisions about resource availability and allocation, the relative priority of system fixes, or how a given change to the system may impact other systems. Management must provide this kind of leadership, and it is required for safety success.

The Working Interface

We have said throughout this book that facilities, equipment, procedures, and behaviors all come together at the point at which worker activity interacts with the technology. We call this point of interaction the Working Interface.

The Working Interface is the heart of the issue in safety performance excellence. It is where injuries will happen or be prevented and it is where safety interventions have their impact. At any given moment, a particular interface is more or less free of exposures. When an interface is rife with exposures, injuries soon follow. The organization that truly wants an injury-free workplace must achieve a Working Interface consistently free of exposure. The task of safety improvement is to engage employees at all levels in exposure reduction through identifying and removing or controlling the hazards in the Working Interface.

The Poison of Blame

Blaming is always counterproductive. When an injury or incident occurs, the important question is not who is at fault, but what can be learned from the incident to prevent it from happening again. This seems obvious, but blaming is an easy trap to fall into: *"If the worker hadn't reached into the paper machine while it was running, he wouldn't have lost his fingers. He's just lucky he didn't lose his hand. It's that simple!"*

Of course, it's not that simple. Blaming the employee may be easy, but it ignores the deeper organizational causes at work. What level of production pressure did the employee feel? Is the design of the equipment optimal? Is reaching into the machine part of the culture — "the way things are done around here"? Do workers regularly reach into moving equipment? Do supervisors overlook it when workers reach into moving equipment?

Blame begets blame. It polarizes.

"If the company did a better job of maintenance and had not created a schedule and production pressure that made shutting down the machine during a run virtually impossible, the worker would not have tried to clear the machine manually."

The oversimplified dichotomy of "who's at fault?" should be rejected at the outset. Blame is contrary to the cooperation required for a comprehensive and effective safety solution. People must *willingly* participate in identifying hazards and implementing solutions.

If the situation is already polarized, the most powerful first step a leader can take is to refuse to blame. Any comprehensive safety solution includes putting an end to blame and instead, focusing on collaborative solutions.

The Leadership Opportunity

Improved safety is of vital interest to both labor and management.

- Unionism has its roots in the need to protect workers. Assuring worker safety is central to the labor leader's interest and one of the imperatives that guides union leaders' actions.

- Management faces exposure to work disruption, direct and indirect costs, and public relations problems. World-class safety performance is a strategic element in world-class performance overall. And organizations have a moral imperative to provide a safe workplace and protect their workers from accidents and injuries.

- *Working together to prevent the kinds of paper machine jams that had historically resulted in amputations, management and workers were not only able to achieve a safer Working Interface and fewer injuries but also a more continuous and reliable production flow.*

Successfully attacking the shared challenge of safety improvement can lay the groundwork for other instances of working together for a common good. Safety is an ideal, mutually advantageous place for labor and management to work together and establish a measure of trust. It may lead to other mutually advantageous cooperative efforts. Thus, both labor and management have a great deal to gain by cooperating to prevent injuries.

This means that safety provides a tremendous opportunity for company and labor leaders. Since safety is strategically important to both, it provides an opportunity to enact a shared core value and objective: concern for the well-being of employees. This is where credibility is made or lost, where valuing employees becomes a reality or is shown to be just talk. Credibility, earned through real action on behalf of employee well-being, is invaluable. And, most importantly, it is simply the right thing to do.

Getting Engaged in Safety

What does it take for employees at all levels to become engaged in the safety effort?

Leadership Vision and Communication

There are many benefits that flow from improved safety but they are not always obvious to every employee. When everyone else feels that the status quo is good enough, the leader must have the vision to see how improvements can be made. The leader is the first to sense the need — and therefore the opportunity — and the first to recognize the path forward.

The leader must be able to make a compelling case for this need and engage others in an ongoing process to identify and remove exposures. Engaging others at all levels is critical to the leader's success as he or she cannot achieve change alone. Others have the energy and knowledge needed to help shape the solution and assure its success. They often know things about the situation that the leader doesn't. Sharing information across the organization is extremely important to the identification and removal of exposures.

The Leaders' Role in Engaging Others

What can formal company leaders do to foster the engagement and communication required of a successful safety process? First, leaders set the direction and create the atmosphere for success.

In addition to avoiding a "blame" culture, company leaders must not tolerate "killing the messenger." Safety-related communication needs to flow easily and quickly from whoever has pertinent information to whoever needs it.

In Chapter 4 we discussed how people's willingness to become engaged and to communicate are a function of cultural attributes. Organizations whose employees perceive lack of fairness (Procedural Justice) and low trust in supervision (Leader-Member Exchange) will not foster employees going above and beyond for safety process improvements.

Employees at each level of the organization need to find ways they can contribute to the safety challenge and come up with ideas about how they can change the systems to achieve a high level of safety performance. If people put their time and energy into something, they want it to reflect their concerns and actually make a difference. Therefore, the things people commit to change must be real and have a substantive and measurable impact on safety performance. They might include such things as:

- Improving the timeliness of responding to identified exposures
- Improving how safety-related information is used when making operational decisions
- Improving compliance with procedures
- Increasing incident reporting
- Improving the incident investigation process

At the worker level, leaders . . .

- Inform and help design improvement plans
- Play key roles during intervention processes
- Take an active role in identifying exposures and making suggestions for how to deal with them

The Safety Improvement Mechanism

A mechanism is a set of steps or system components that reliably lead to a defined result. Improving safety continuously, rather than in fits and starts, requires establishing ongoing mechanisms that become normal practice and are supported by the organization's culture. Those described in this chapter are designed to produce fewer injuries. What we describe should not be misconstrued as a new safety program because safety programs can come and go; the mechanism we seek to create is ongoing.

Five integrated, systematic sub-processes make up the core mechanism that reduces exposure to hazards:

- **Identify Exposures** to hazards

- **Gather Leading Indicator Data** on the occurrence of exposures.

- **Use Data to Provide Feedback** to the organization about the status of exposures and their removal

- **Reduce Exposures** based on leading indicator data

- **Renew the Processes** based on near miss and injury experience

Many companies use one or more of these processes already, but unless they are integrated and deployed systematically, they don't create a reliable mechanism and the results they deliver will be inconsistent and sub-optimal.

Although these sub-processes can be implemented in many different ways, they all focus on the hazards in the Working Interface and the ways in which employees become exposed to them. Each of the five processes needs to be implemented to maximize the engagement opportunities for people at all levels of the organization. What this looks like in actual practice is a function of the circumstances in which they are implemented.

Identify Exposures

Exposure reduction is a value for all employees. Thus, if they expect that action will be taken, the opportunity to identify exposures will motivate them to become engaged. The design of the safety improvement process needs to provide many opportunities for this to occur.

- Front-line employees can work with supervisors, managers, and safety professionals to identify exposures.

- Supervisors and managers can investigate how well their systems reveal exposures or inform them of any new exposures created by their operational decisions.

- Root cause analysis can be completed for all injuries and near misses that have occurred over a recent period. Exposures in the Working Interface can be identified based on this analysis and then classified by category. Operational definitions can then be written to spell out what constitutes a safe Working Interface.

Gather Data

Gathering hard data about exposure to hazards provides information needed to: 1) provide leading indicators of ongoing performance; and 2) identify specific areas for improvement. Data on the frequency of exposures will inform and justify resource allocation. Gathering data for this purpose can be tremendously motivating because it is a tangible way to make a difference.

Data gathering can be done effectively in a variety of ways. Some alternatives include examining work processes, interviewing people involved in the work about exposures, or observing work in progress. This doesn't necessarily mean doing safety observations; observation is one way to gather data, but not the only way. There is no one "right" way to gather data on the extent of exposure to hazards. When choosing a data gathering method, consider the nature of the existing culture and how workers will perceive the particular method under consideration.

Two more examples of data gathering methods are tracking safety suggestions and action items. And in some office applications, in response to a periodic email prompt, workers gather ergonomic information by self-report and a standardized workstation evaluation.

Use Data to Provide Feedback

High-performing organizations are feedback-rich environments. They have mechanisms that focus this feedback on exposure-recognition and -removal activities. They explicitly recognize the value of these activities to the organization and work group. This supports the value of safety in the organizational culture.

It's important for the organization to know, in specific quantified terms, how it is doing on the critically important task of improving the Working Interface. Since this information is a leading indicator, it is also important for workers and leaders to see it, as a measure of progress and a call to action. Seeing progress in this area is highly reinforcing to employees at all levels.

Use Data to Reduce Exposures

Once exposures have been identified, leadership can now take action. They can set improvement targets by analyzing the data and prioritizing based on frequency and severity of injuries potentially associated with each exposure.

Using the Hierarchy of Controls • Specifically, this sub-process entails reducing exposure by using the control strategy as close to the top of the hierarchy as possible and building in as much redundancy as the frequency and severity potential warrant by adding additional lower levels of control as needed: For example:

- **Elimination:** Removing the hazard at its source
- **Substitution:** Replacing a procedure or chemical with a less hazardous one
- **Engineering:** Installing guards or interlocks on machinery
- **Administration:** Implementing safety policies and procedures
- **Personal:** Using protective equipment, watching out for co-workers

Workers can provide valuable input for higher-level interventions, and their buy-in and cooperation will strengthen the improvement process.

The exposure reduction processes usually make use of the organization's pre-existing problem-solving mechanisms. These may need to be modified to accept input from exposure data and sometimes from workers. Ideally, problems that can be solved by workers and supervisors should be. Those that require commitment of more resources or have a systemic basis require more formal processes.

After implementing a plan to reduce or eliminate an exposure, the data-gathering process should confirm that the plan worked. This step is often overlooked. Because they involve the interaction of the worker with the technology, exposures are rarely static. Therefore, common sense predictions need to be tested to assure that the exposure is gone and that the Working Interface is safe.

Renew the Process

Renewal has two aspects:

1. Evaluating the ongoing relevance of the improvement process to the specific injuries and near misses that continue to occur. This amounts to re-calibrating the safety mechanisms against current exposures. An outline is shown below:

- Examine the root causes of each incident and near miss to find the exposure responsible. Examine exposure data to determine whether this exposure had been previously identified. Assure that data gathering includes the exposures identified in incidents and near misses.

- Examine the relevance and impact of safety improvement efforts and enabling systems on current exposures.

- Determine whether any of the sub-processes need to be adjusted to increase or sustain the safety improvement process' impact? For example, have there been significant changes in the configuration of the Working Interface that require new exposures to be identified?

2. Organizations are not static. Now, perhaps more than ever before, change is a way of life. Objectives, procedures, schedules, equipment, and workers change over time, sometimes rapidly. This means that the people, processes, and challenges present when the sub-processes were implemented may change in months or years, and that periodic review is necessary to assure optimal functioning. Leaders need to be engaged in a renewal process to update each of the other four processes that make up the safety improvement mechanism.

Implementing the Process

As is true of everything else in safety improvement, the implementation process provides a tremendous opportunity to foster across-the-organization engagement. Employees at all levels and positions get a chance to demonstrate safety leadership.

Implementation Team Makeup

Under most circumstances a team is created to implement the process of systematically reducing hazards. The most effective implementation team is composed of a cross-section of employees as well as a management sponsor who will champion the process with the leadership group. If there is a union, a comparable leadership sponsor should be added. The individuals representing labor and management should have a good working relationship or be willing to develop one.

Team members should be natural leaders who are trusted and respected by their peers and represent a cross-section of the organization. They should be selected for their ability to contribute key talents, their dedication to safety, and their enthusiasm for the process. Useful talents and background experience include:

- An understanding of the organization
- Time management skills
- Communications skills, including public speaking
- Effectiveness at training
- Computer skills
- People skills
- Task orientation
- Team-building experience

Obviously, no one person can bring all of this to the table, but these skills should be represented on the team as a whole. Although the team should represent all key groups in the organization, it must also be small enough to be an effective working group, usually not more than 12.

If the organization has more than 500 employees or is widely dispersed, more than one team will likely be required. In this case, the individual or group to whom the teams report (usually the senior management team) should provide a coordinating mechanism for them.

Implementation Team Charter

The work of the team includes:

- Organizing and participating in the assessment of the current state
- Communicating assessment findings to the organization
- Designing and implementing the five processes that constitute the hazard reduction mechanism
- Communicating and obtaining buy-in for the process

- Implementing each of the processes. In addition to the specific steps in each process, this includes:

  — *Recruiting others to participate in the various processes*

  — *Developing and performing the appropriate training*

  — *Overseeing the quality of the process and providing improvement interventions as needed*

  — *Coordinating and integrating the processes with each other and with other safety systems*

- Measuring the health of the processes

- Reporting to management, labor, and the organization at large on progress, issues, and barriers

- Working with management and labor to overcome barriers

While the implementation team plays a central role in the process, everyone else in the organization has a role as well, and these roles need to be defined with input from the individuals involved.

Roles at Every Level

Workers' Roles

Becoming actively engaged in injury prevention is desirable for every worker. Workers can become more formally engaged by taking on special tasks — from presenting at a safety meeting to working on a safety problem-solving team — or by taking on special roles such as becoming a member of the implementation team. Whatever form their engagement takes, it needs to be fostered and recognized by supervisors.

Supervisors' Roles

The foundation upon which the supervisor builds leadership is the relationship he or she establishes with reports (Chapter 8). This relationship must be authentic and build toward strengthening any areas of low organizational functioning (Chapter 4).

Some of the most critical things supervisors must do to support the process follow:

- Avoid blame and reinforce even small efforts at upward safety communication

- Provide the time and resources needed by the implementation team members and others who are active in the process; backfill to avoid a backlash against them or the process due to their needing to be away from the job

- Follow through on safety commitments in a timely way

- Provide frequent feedback on the status of hazard reduction efforts

- Explicitly include safety considerations in all operational decisions

- Encourage, recognize, and reinforce the safety activities of workers

- Communicate effectively — both upwards and to reports — on behalf of safety

- Participate in more specific roles as necessary

Managers' Roles

Accountability and performance mechanisms need to focus on reinforcing managers for their implementation successes. Other important aspects of management's role include:

- Receive input from the implementation team and others about the health of the process and work with them to strengthen it

- Monitor and reinforce supervisor performance vis-à-vis their role in the process

- Positively reinforce upward communication for safety

Leadership and Reduction of Exposure to Hazards

Our research shows that successful processes have high-quality leadership and management support. Everything leaders do for safety is aimed at facilitating whole-hearted, willing safety engagement at every level in the organization. It is this engagement that produces safety results and assures sustainability. In one sense, sustained engagement *is* culture change. This is what leaders at all levels are working together to accomplish.

Preparing the Soil

Reducing exposure to hazards is an organizational development undertaking. It can be thought of much like constructing a garden. The first step is to prepare the soil in which the garden will thrive.

The organization's senior-most leader must take the initial step in this preparation. He or she must sponsor the initiative to improve safety. This means insisting on both the urgency of the challenge and opportunity, and on the broad shape of the solution. Setting such direction is not a leadership function that can be successfully delegated.

Sometimes, the site leader may be unable to sponsor the undertaking without the explicit support of those higher in the organization. Even if he or she can, it is imprudent to do so. He needs their commitment. Their role in the process is to back him up on the direction he wants to take his site. And he should make it easy for them to execute their role by keeping them informed on progress and results. In many organizations, the site manager makes members of the implementation team visible to corporate safety meetings, both as a perk for their work and as a means of sustaining interest and support at higher levels.

Next, the senior leader must ensure that all of the other senior-most leaders — both organizational and labor leaders — are aligned with him on the undertaking. Such alignment must not be simply assumed. Often these other leaders will have serious reservations. Sometimes they may be reluctant to put their reservations on the table, but they must or their concerns will not be addressed as the process moves forward.

As a start, the alignment process must involve clarifying the current state, the desired state, and the gaps. It then needs to move on to action planning and accountability. In the end, all leaders must share a common view of the following:

- What they are aiming to accomplish and why
- How they expect to get there
- What the process will require of them individually
- How their performance will be measured
- How they will know they are successful

Best Practices

To understand what the process requires of them, leaders will have to identify and define the specific behaviors they need to perform in each of the seven leadership best practice categories discussed in Chapter 3. These will include things they need to change about how they manage now and tools they will use for supporting the safety process. Chapter 3 discusses these practices and how they can be used to build a climate in which the safety process can grow and flourish.

These practices will be deployed in support of each of the five sub-processes that constitute the safety improvement mechanism. Each of these component processes provides opportunities for leadership to demonstrate its support. The "vision," for example, should include all of these processes. Below are some additional examples:

- **Identify Exposures:** This demonstrates *action-orientation;* it can be done in ways that demonstrate *communication, collaboration,* and *recognition* as well

- **Gather Data**: Fact-based decision-making builds *credibility;* again, it can be done in ways that also demonstrate *communication, collaboration,* and *recognition*

- **Provide Feedback:** Using exposure data to provide performance *feedback* creates a feedback-rich environment and *accountability*

- **Control Hazards:** *Taking action* in a *collaborative* way in response to information about hazards, working to remove them, and communicating results builds *credibility.*

- **Renew the Process:** Renewing requires *vision, commitment* to the long haul, and *action orientation*. It builds *credibility*. To renew effectively, both requires and demonstrates *collaboration, communication, feedback and recognition, and accountability*.

Prepared with their best practices and a plan to implement them throughout their own parts of the organization, the leaders are ready to initiate the implementation of the safety improvement process.

Getting Started

The first step in evaluating readiness is for the leader to ask himself why he or she wants to implement a safety improvement process. How does it relate to the organization's strategic objectives?

- *What value do I expect it to deliver?*
- *Who else in the company would see this as valuable? Who would not?*
- *How much time do I have for the safety improvement process to deliver this value?*
- *What will be the consequences to me, and the organization, if I don't implement a safety improvement process and fail to achieve the value above?*

These considerations will create awareness of the importance of what he wants to accomplish as well as point out the constraints within which he is operating. All he may need for now is a successful demonstration project. Or his primary objective may be substantial culture change. He may have a lot of time — 12 months or even several years — or he may need very quick results.

There are many approaches to implementing safety improvement processes. The actual approach must be appropriate to the objectives, the constraints, and the resources available. Which of the component processes are most critical to the objectives? Could they be implemented successfully as stand-alone processes? Could they be the first step in a phased, full-scale implementation? Or is it better to bite off the whole thing at once?

Diagnosing the Current State

Once the leader has formulated some ideas, he should then consider the current state of things to get a clear handle on the strengths and weakness of the culture in which changes are to be made. Assessing the organization's culture is a particularly important step if the plan is to implement a complete process throughout the entire organization. The Organizational Culture Diagnostic Instrument or a similar instrument can help with this question and often stimulates positive interest for change (Chapter 4). Taking the OCDI and responding to its findings may deliver all the information he needs.

Getting Started with Leadership

If the process needs to start slowly, important early steps often involve limiting the work to developing understanding and support among other leaders. This can include such things as:

- Training presentations
- Workshops in which leaders either work on their safety vision and alignment or do other pre-work needed to plan an effective approach
- One-on-one executive coaching to develop safety leadership skills and abilities needed to lead and manage an organizational development/culture change effort (See Chapter 7)

Getting Started with Supervisors

Alternatively, sometimes the best place to start is with supervisors. The greatest opportunity for quick impact may lie in developing supervisory people skills and directing their new skills at building the kind of relationships that support a healthy climate. Again, making this decision is a matter of establishing a solid up-front diagnosis of the current state.

Summary

Important things to consider in evaluating readiness to implement a new safety improvement initiative and shape a path forward include the following:

- How closely is safety related to the organization's strategic objectives?
- How clear and well articulated is leadership's safety vision?
- How well aligned are the senior leaders?
- How much support for their vision exists at various levels in the organization?
- How skilled are leaders and managers vis-à-vis organizational development undertakings?
- How skilled are supervisors in sustaining the right kind of relationships with workers?
- How able are union and management leaders to cooperate over a common objective?
- How much resource is available for the undertaking?
- Which levels or groups in the organization have the most time available to devote to the undertaking?
- Who are potential champions, supporters, and blockers?

What to Expect

When people at all levels of the organization...

- Work together cooperatively and with sufficiently open communication;
- Identify, measure, give and take feedback about exposures, and respond effectively to workplace hazards;
- Continuously improve their safety performance; and
- Interact in ways that raise the organization's OCDI scores...

They will have successfully moved the organization a long way down the path of building an injury-free culture. Along the way they will have prevented many injuries and installed a proactive, continuously improving safety process.

• • • • •

Section 5

Applications

With a solid set of tools, an understanding of crucial human performance issues, and a set of methods to create engagement, we are ready to design specific interventions to achieve safety improvement. We can draw on everything learned up to now to craft the solutions best suited for the particular needs and objectives of our organization. In this section we start with an organized way to assess needs and design intervention plans, and then look at some specific case histories.

Chapter 10

Planning for Change:
Designing Intervention Strategies
for Safety Improvement

This chapter was written in collaboration
with John Hidley, M.D.

- The importance of having an effective strategy
  for safety improvement

- Developing a strategic plan for safety improvement

- Examples of the development of strategic plans for safety improvement
    Armed services branch
    International metals and mining company
    International energy and utilities company
    Gulf coast chemical producer
    Puerto Rican consumer products company

In this chapter we will address how to develop an effective strategic plan for safety improvement. Where strategy development ends and execution begins is not always a clear and bright line. Looking back at successful change initiatives, leaders often notice that the change really started at the point at which they clearly understood the current state and the desired future state. Nevertheless, it is useful to review strategy development and execution as separate elements.

The Critical Importance of Having an Effective Strategy for Safety Improvement

We all know the importance of strategy to good business outcomes. We've attended workshops and read books and articles about it, and we've been part of efforts to improve our organizations' ability to find and execute effective strategies. So, it comes as something of a surprise that many companies we have worked with in safety improvement often fail to realize they need a strategy for safety improvement. Senior leaders recognize that it must improve and look for ways to make improvements, but they often fail to develop a strategic plan, or they wait until resources have already been used ineffectively and leverage has been unnecessarily lost.

For our purposes a strategic plan consists of an integrated plan of action aimed at achieving a set of objectives. In the absence of a clearly articulated strategy, understood well and supported fully by the leadership team, the likelihood of successful execution is greatly reduced.

The strategic plan for safety improvement should include the following elements:

1. A set of operationally-defined objectives that can be summarized by a short vision statement

2. An assessment of the current state that defines specific gaps between it and the desired future state

3. An intervention plan that addresses the gaps and names accountabilities

4. Critical success factors that specify the ways things must be done to be successful

5. Measures to monitor the change process

The planning process does not end with the development of the plan. As the plan is put into action, it is often necessary to make revisions and adjustments as new information and insights are discovered.

Developing the strategic plan for safety improvement

We will give examples of strategic plans for safety improvement later in this chapter. First, we'll outline the core elements of an effective strategic plan for safety improvement.

Objectives and Vision Statement

What is it the organization really wants to accomplish in safety improvement? Often organizations think of this too narrowly. "We want to move from our current corporate-wide recordable incident rate of 2.1 to below 1.0." "We want to create an injury-free culture." "We want to stop having fatalities."

These are fine objectives, but they leave a lot unsaid. How important is the longevity of the change that is achieved? Does the organization want to reduce incident frequency and have confidence it will stay low and get better, or is it unconcerned about a rebound effect? Have previous initiatives been perceived as temporary? How important is this to the organization's culture? Is it important that the safety initiative complement other initiatives already underway? If so, what needs to be considered to ensure appropriate integration of initiatives? Is the initiative primarily about safety improvement, with leadership improvement as a tool and fortunate side effect, or is the primary objective leadership improvement, with safety improvement as a fortunate side effect or entry point? Is the culture change wanted to support the safety effort or is culture change pivotal to performance generally?

In addition to the summary statement of objectives or vision, a specific statement should address these issues by defining the desired future state. This should have a brief summary — for example, "We want to be the best in our industry" or "We want to be world class in safety performance" or "We want an injury-free culture." But it should also have a few paragraphs that define the future state in operational terms. Often this is not totally clear to leaders at the outset, and only becomes clear as gap analysis is completed.

Assessment of the Gap
Between Current and Desired State

A vision sets the stage in broad terms, but to have clarity about the desired future state it's necessary to understand the current state. Without this critical information, the organization will inevitably do things that are unnecessary and address the wrong objectives, underestimating the need for some changes and overestimating the need for others.

How will the current state be assessed? The Organizational Safety Model introduced in Chapter 1 provides an excellent framework for thinking about the current state. (Figure 10-1)

In Chapter 1 we discussed this model in detail, so we will only summarize it here. Improving safety means eliminating hazards where the worker and the technology interact, which we call the Working Interface. Safety enabling systems are the core safety programs and processes the organization uses to control exposure to hazards. Sustaining systems are the organizational systems that provide support and longevity to safety systems. Culture is the values that drive the organization and set the tone for safety, the emphasis given to safety, unstated assumptions about how things are done. Leadership is the force that energizes everything else in the model. It creates the culture, ensures that systems of both kinds are in place, and ultimately regulates the Working Interface.

Most organizations can see themselves pretty quickly when looking at the Organizational Safety Model, at the macro level. You know you have the safety infrastructure in place but your sustaining systems to support it are not as strong as you'd like. The culture is strong in some parts of the organization but not others. And leadership is dedicated but not performing

Figure 10-1. The Organizational Safety Model.

consistently at the level of best practices. That's the easy part.

The more important part is to quantify each aspect of the model for each level of the organization. This is critically important for designing an effective intervention plan. Otherwise, a one-size-fits-all approach will result in wasted resources and less than optimal outcomes.

The tools described in this book can be used to assess the effectiveness of leadership, culture, and aspects of the model. Leadership effectiveness is assessed using the Leadership Diagnostic Instrument (Chapter 3) and profiles of the Big Five personality dimensions (Chapter 2). Culture is assessed using the Organizational Culture Diagnostic Instrument (Chapter 4), and the Working Interface is assessed using employee-driven methods to identify exposure to hazards (Chapter 9).

Assessment Tools		
Leadership Effectiveness	**Culture**	**Working Interface**
Leadership Diagnostic Instrument	Organizational Culture Diagnostic Instrument	Employee-Driven Methods of Exposure Identification
Personality Profiles		

Table 10-1. Assessment tools.

Enabling safety systems can be measured by audit systems, and sustaining organizational systems are measured by interviews and surveys of employees, as well as reviews. The leader can then examine how safety-enabling systems and sustaining systems, as well as culture and leadership, are impacting the interface.

Using multiple assessment tools is highly desirable because it allows for cross-validation of findings. When interviews and survey diagnostics both find the same things, confidence in the accuracy of the assessment goes up.

Having good data on each aspect of the model enables the strategy team to identify gaps and begin to formulate plans. The following considerations should be taken into account when putting the plan together:

Leverage: some potential targets will have greater impact than others. For example, building a strong safety culture can have deep and long-lasting effects. Strengthening safety leadership capacity has perhaps the greatest potential for impact. But leadership, culture, systems, and the Working Interface all interact; they are a system. Focusing exclusively on one of this system's elements when there are significant gaps in other elements can be self-defeating.

Breadth: how extensive should the intervention be? Should it target just those elements of culture that influence safety, or the organizational culture more generally? Should it aim to improve leadership generally, or zero in on safety leadership? Can it be more narrowly focused on specific enabling or sustaining systems?

Organizational Scope: does the entire organization need to be included, or only parts of it? At what level in the organization is attention most needed — e.g., site, division, or corporate? Are there specific sites or divisions that require attention?

Urgency and Time Requirements: how fast are results required? Do the needed changes require a long- or short-term effort? If urgency is high and the required changes are difficult or extensive, consider developing both a short-term and long-range plan.

Measures: how can a baseline be established? How can progress be tracked? It is important to consider both the outcome and up-stream, process measures. Costs and benefits of the initiative should also be tracked.

Intervention Plan

Given the gaps between the current state and the desired future state that the assessment has revealed, we can now craft an intervention plan. Organizations do this in different ways. Some use their own internal resources, while others use outside resources, or a combination. The tools we have found to be most effective fall into four categories. Each category is applicable at the corporate, division, or site level.

Organization-wide strategic change initiative. This kind of initiative is driven from the top and requires the active support and participation of the organization's senior-most leader. It is broadly strategic and ties safety improvement objectives to other organization-wide objectives. We'll describe several examples of this kind of initiative in this chapter and Chapter 11, as well as a case study of NASA, which comprises all of Chapter 12.

Leadership team initiatives. This is typically done for the senior-most leadership team first and then filtered down through the organization as needed. These initiatives generally involve a combination of workshops (which vary from one to three days), individual leadership feedback assessments for each participant, and/or some group or one–to-one coaching.

Employee-driven safety initiatives that target improvement of the Working Interface. These processes can be done at the site level as well as on a division- or corporate-wide basis. The mechanisms that make these initiatives effective are employee engagement and leadership support. How these processes are implemented varies widely depending on the overall situation found in assessment of the gap between current and desired state.

Supervisory-development initiatives. In many organizations the assessment process reveals that a pivotal area requiring improvement is lower-level supervisors' leadership skills. Usually these initiatives include an assessment of the leadership strength of the individual and then a combination of skills training, coaching, and feedback, either on a group or one-to-one basis.

The particular intervention plan for a given organization should not be taken from a template or a set of tools. Rather it should be crafted, based on the findings of the assessment process, and the objectives and resources available.

Balancing Objectives and Resources

The need to find the right balance between objectives and resources underlies all of the activities of defining objectives and building intervention plans. The question that must be considered is "Have we planned adequate resources to reach the objectives outlined?"

Lack of resources is perhaps the most frequent cause for failure we have seen in improvement initiatives. For example, a one-day training workshop for leaders, no matter how effectively conducted, will not change the culture of the organization, but it may set the stage for change, create a shared vision among leaders, and begin a process that will be successful. A two-day training course for site-level supervisors will not modify the Working Interface permanently, but it may enable supervisors with a set of tools, a way to think about influencing the Working Interface, and a new level of motivation.

Critical Success Factors

Knowing your organization, its strengths and weaknesses, history of change initiatives, and business demands, what must you do to be successful with the intervention plan you have crafted? These critical success factors should be identified and leadership needs to gain alignment around their importance.

We'll give specific examples below, but the general idea is to consider first how the initiative is to be implemented, and then to ask where you are likely to overlook, pass by, or not reach far enough to accomplish your objectives.

Examples of the development of strategic plans for safety improvement

In this section we will give several examples of organizations that developed strategic plans for safety improvement using the tools described in this book.

Example 1: Organization-Wide Safety Culture Change in an Armed Services Branch

The senior-most leadership of a branch of the armed services was concerned because of the high number of off-duty motor vehicle fatalities. The senior leaders' decision to do something about this was primarily an ethical decision based on their intrinsic values: they were very concerned for the well-being of their young warriors.

Objectives and Vision Statement

The organization's directive was to reduce off-duty fatalities by 50% in a two-year period. They knew that to do this they must change the organizational culture and safety climate, but that this must be done without weakening their highly successful fighting ethos or unduly burdening their organization during a time of war. The objective was to create a culture in which driving safety was a core value shared by personnel at all levels of the organization. Selected measures were the Organizational Culture Diagnostic Instrument (OCDI) scales, and fatality rates.

Assessing the Current State and Defining Gaps

The current-state assessment consisted of one-on-one interviews throughout the hierarchy of the command structure, focus groups at various levels throughout the organization, examination of enabling and facilitating systems, and evaluation of the organizational culture using the OCDI.

Interviews and focus groups revealed that leadership effectiveness was perceived to be very strong in general, but that safety leadership specifically was in need of improvement, especially at the lowest levels of the organization. In particular, interviews suggested that the willingness to communicate safety concerns was an improvement opportunity that would strengthen the safety climate.

The systems review revealed weaknesses in both enabling safety systems and sustaining systems that were contributing to the perception that safety was not an organizational value. These were also making it very difficult for the organization to practice safety efficiently and effectively. Finally, they impaired leadership's ability to know where the organization stood, the meaning of its accident rates, and the measurement of its progress. For example, "Class A" accidents, those involving loss of life or over $1 million in damage, were the primary measure of safety performance. This data was accurate, though sometimes misleading due to the methods of measurement.

Using the Organizational Culture
Diagnostic Instrument (OCDI)

Impressions from interviews and focus groups were useful, but they didn't provide a way of quantifying specific aspects of the organization's culture. This is critically important for several reasons: cross-validation, specific measurement within operating units, and outcome measures.

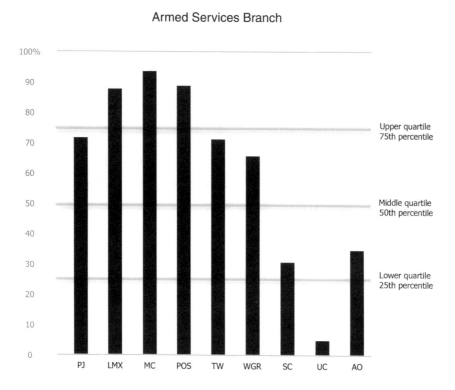

Armed Services Branch

Organizational Dimension	Team Dimension	Safety-Specific Dimension

PJ   - Procedural Justice
LMX - Leader-Member Exchange
MC   - Management Credibility
POS - Perceived Organizational Support
	TW - Teamwork
	WGR - Workgroup
	      Relations
		SC - Safety Climate
		UC - Upward Communication
		      About Safety
		AO - Approaching Others About Safety

Figure 10-2. OCDI scores for all active duty personnel in an armed services branch. Scores are compared to a norms database of other organizations.

The OCDI was given to a cross-section of 1,268 members of the organization. The results were very interesting and useful in helping us understand the organization's strengths and what it faced in making the desired changes.

When reviewing OCDI data, keep in mind that it is reported in percentile terms in relation to a database of several hundred organizations. A score of 90 means the organization being assessed scores higher than 90% of the organizations in the database.

The OCDI is described in detail in Chapter 4, so we will give only a summary explanation here. The leftmost four bars represent the four scales making up the Organization Dimension. The Organization Dimension comprises Procedural Justice, Leader-Member Exchange, Management Credibility, and Perceived Organizational Support. These scales make up the four pillars of organizational culture. They are a direct reflection of the way employees perceive leadership in the organization.

The next two bars report on the strength of the organization's level of Teamwork and Workgroup Relations. The next three bars report on the strength of three aspects of organizational safety: Safety Climate, Upward Communication, and Approaching Others.

This interesting OCDI profile reveals that the organization has strong leadership generally (although the Procedural Justice scale is lower than the other three scales in this dimension), and the team scales are well above average. However, the safety scales are all quite low, and Upward Communication is very low.

This is an interesting profile for several reasons. The Organization Dimension often drives the Safety Dimension, but in this case it doesn't. This suggests a culture in which leadership itself is very strong, teamwork is high, and safety is not emphasized.

This is not surprising for a military organization that values effective combat, essential to its core objectives, but it is incompatible with a parallel need for safety. Given this apparent conflict, it isn't difficult to understand how young warriors, successful in combat, would overlook safety considerations when driving personal motor vehicles off duty.

But the leaders of this organization do not want to see their young members killed in crashes. So they are faced with a serious issue: how to maintain the effectiveness of their warrior culture without sacrificing individuals as a side effect. Although our task was focused on mishap prevention, the organization also saw Teamwork scores as lower than expected. Teamwork is essential to their effectiveness in combat and a key component of their ethos.

Intervention Plan

The organizational leadership embraced these findings and set out very aggressively to develop an intervention plan to address them. They concluded that the issues identified in the OCDI and interviews were service-wide and required an intervention plan that addressed culture and systems throughout the entire organization. They formed three committees:

1. The Safety Culture and Implementation Steering Team to define the ideal safety culture and oversee the entire project

2. The Safety Systems Design and Implementation Team to select and plan an attack on those system issues with the greatest potential for impact

3. The Behavioral Safety Design and Implementation Team to develop a behavioral change plan applicable to all levels

The first committee, the culture change committee, was chaired by the second-most senior leader of the service and had the direct buy-in and support of the senior-most leader. The other two committees reported to the culture change committee, which reported to the Executive Safety Board comprising all of the senior-most leaders of the service.

These committees developed detailed intervention plans to significantly improve each of the three areas. Because of the complexity of the situation, the plans involved multiple timelines. For example, the timeline for improving the safety culture was longer than that for making sustaining systems improvements, for example, building safety performance into advancement criteria.

The process involved ensuring that the service took ownership of the plan and that it aligned with its culture and values. The teams decided that leveraging their strength in leadership would produce many desired results — of which safety is the bellwether.

Critical Success Factors

The following three factors were identified as critical to the success of the undertaking:

- Provision for the rapid rate of turnover at the top of the organization to assure continuity and support across time

- Effective inclusion of new organizational/cultural mandates in the leadership process itself and in all other training programs, starting at recruitment

- Timely implementation of effective process-and-outcome measures to track progress and success

Example 2: Corporate-Wide Safety Leadership Development for an International Metals and Mining Company

This international metals and mining company recognized midway through an internally supported corporate-wide safety improvement initiative that leadership best practices were not being utilized and that they were needed to reach the company's objectives. The organization employs about 100,000, and has sales of $25 billion in five business units.

Its new CEO, whose background was in the process industries, was shocked by the organization's safety performance, particularly the number of fatalities. He saw safety improvement as a strategic direction for the company. His motive was in part ethical — he said he would not tolerate anything less than world-class safety in the company's operations — but it was also financial and strategic. He believed that safety was an ideal vehicle for leadership development.

Objectives and Vision Statement

"Create the leadership capability needed to support an injury-free culture and in the process strengthen the organization's leadership bench strength. This will require developing new motivations and skills in leaders at all levels."

Assessing the Current State and Defining Gaps

The Leadership Diagnostic Instrument was given to all senior leaders above the plant manager level. The instrument takes a 360-degree measure of each leader's style as well as his or her performance on safety leadership best practices (Chapter 3). Leaders were then provided with a report summarizing their individual results and showing how they ranked compared to leaders in other organizations. Here we will describe the summary report briefly.

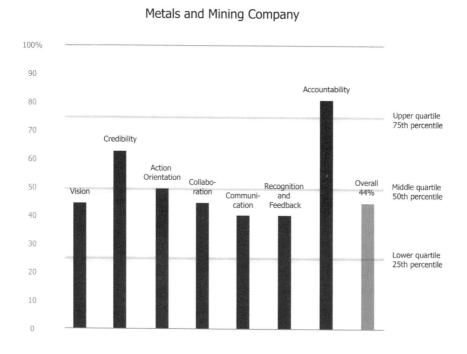

Figure 10-3. Combined leadership best practices scores for metals and mining company. Scores are expressed as percentiles showing comparison to other organizations.

Figure 10-3 shows the best practices results for the group of leaders of one of the company's organizations. Again, these scales are in percentiles against a norms database. As you can see, the leaders as a group are quite strong on practices related to Accountability and Credibility, but they are weaker on the practices that have communication as their basis: Recognition and Feedback, Communication, Collaboration, and Visionary practices are all below the 50th percentile.

These data are typical of this organization. They pinpoint the gap between the current and desired future state, and show that the focus of the intervention plan needs to be on motivating leaders around the need for safety, educating them about what they can do for safety, and helping them develop the critical "soft skills" — the communication practices — to make it happen.

Intervention Plan

We developed a leadership training seminar for all corporate leaders above the plant manager level. The seminar addressed both the motivational aspects of safety, as well as more practical, hands-on issues. Each individual leader received a Leadership Diagnostic Instrument like that shown above. Each leader also received two hours of leadership coaching on devising a safety improvement plan. Additional coaching was made available as requested. In addition, each leader was given the option of cascading the process down into the organization.

Critical Success Factors

The following factors were critical to the success of this plan:

- Continuing emphasis, support, and active participation by senior-most leadership

- A mechanism to monitor and reinforce the success of the developmental action plans

- A mechanism to monitor the impact of the plans on the leader's organization

Example 3:   Integrated Safety Leadership Development for an International Energy and Utility Company

This organization is one of five major business units in an international billion-dollar energy and utility company. As part of its strategic plan, the company's leaders aspired to make the company a world-class safety performer.

It had previously implemented a sophisticated leadership development process, which had been well received, but safety leadership had not been fully embraced. Although this process was successful in some ways, it had not been fully integrated at the day-to-day level, and the division head was searching for additional ways to improve the organization's leadership. Safety improvement was seen as an opportunity worth pursuing in its own right and also one that would tie in with needed leadership improvement generally. All of the organization's refineries had either implemented an employee-driven safety process or were in the process of doing so.

Objectives and Vision Statement

"Become world-class in safety performance. In contrast to our industry, have incident frequency rates in the first quartile or better. This will require measurably improved leadership capability throughout the organization and installing an improved process of ongoing leadership development. This process must be successfully embedded in our ongoing operation. All of this must strengthen our employee-driven safety initiatives by additional leadership involvement, increasing their sustainability and making them a part of the day-to-day work culture."

Results will be measured by repeated administrations of the OCDI, Leadership Diagnostic Instrument, progress on internal leadership performance ratings, and incident rates.

Assessing the Current State and Defining Gaps

We administered Leadership Diagnostic Instruments to all leaders, from the refinery leadership teams to the senior-most leadership team. The OCDI was given to all locations. Figure 10-4 below shows the OCDI scores for one of the refineries and Figure 10-5 shows compiled scores on the Leadership Diagnostic Instrument.

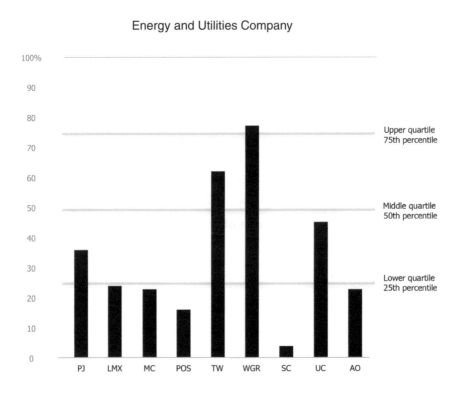

Figure 10-4. OCDI scores for one refinery at energy and utilities company. Scores are expressed as percentiles showing comparison to other organizations.

These scores reveal that team functioning and Upward Communication are relatively strong (within this profile). But the other scales are low. Employees perceive leadership as distant, and the Safety Climate in the organization is low. This suggests that leadership and supervision both need to get much more involved in the safety effort.

The combined leadership best practices scores are consistent with this interpretation:

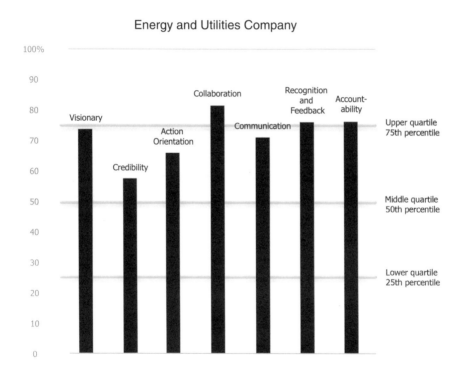

Figure 10-5. Combined leadership best practices scores for energy and utilities company. Scores are expressed as percentiles showing comparison to other organizations.

Although there are individuals who are exceptional leaders, the combined scores show that leadership overall is relatively unexceptional in all areas of critical safety leadership best practices.

The consultant worked with the leadership team to refine what had been learned from implementation of previous initiatives and conducted additional interviews to flesh out needed information about the current state.

Implementation Plan

The gap was clearly with leadership, and the intervention plan used multiple strategies to focus on leadership development at both the visionary and hands-on levels. Elements of the intervention included:

- Interlocking, cascading leadership workshops
- Supervisory development workshops and action planning
- One-on-one leadership coaching
- An observation-and-feedback process for leadership development
- Integration of leadership behaviors into work contracts

Leadership workshops

We developed a set of interlocking leadership workshops and gave them starting at the top and cascading down. Workshop participants constructed individual development plans based on their own survey results and then received one-on-one coaching to help them most efficiently make the changes the survey indicated they needed to make.

Behavioral Observation and Feedback for Leadership

Young fast-track leaders were enlisted to work with senior leaders to develop an inventory of behaviors critical to leadership success in this company. Senior leaders were trained and coached to model these behaviors. The younger leaders were then trained to observe the behaviors and provide feedback to the more senior leaders. In this way the young leaders gained real understanding of leadership by having their attention focused on critical leadership behaviors. The senior leaders received positive feedback for being good leadership models. The critical leadership behaviors were integrated into all leaders' performance goals and the leadership observation.

Supervisory Development Methodology

At the supervisor level, workshops focused more on building supervisory skills, and the supervisors' action plans focused on implementing their new skills in their day-to-day work. Their action plans became part of their work contracts, and their managers followed up on their progress.

Critical Success Factors

Success factors in this project included the following:

- Maintaining focus. This was especially important since the undertaking was so large and diverse

- Successful integration with existing leadership development processes

- Ongoing commitment and active involvement of senior leadership

- Willingness of senior leadership to listen sensitively to the pressures the sites were experiencing and be flexible, while at the same time, not sacrificing the core importance of successful leadership development

Results of these initiatives will be discussed in Chapter 11.

Example 4: Site Level Culture Change and Safety Leadership Development for a Gulf Coast Chemical Producer

This 500-employee Gulf Coast site of a major international chemical producer had implemented an employee-driven safety process in the late 1980s and enjoyed a trend of decreasing incident rates, which had leveled at a recordable rate slightly above 1.0. However, over 15 years its safety process had grown stale and the corporate parent set a new objective that all locations operate below a recordable rate of 0.5.

Objectives and Vision Statement

"Reduce recordable incident frequency to below 0.5. Engage leadership at each level in the improvement process in a way that drives safety culture change and sustainable change."

Measures of success will be repeated administrations of the OCDI, leading indicator data from the employee-driven process, and incident frequency rates.

Assessment of Current State and Gap Definition

OCDI data are shown in Figure 10-6. The OCDI profile shows high levels

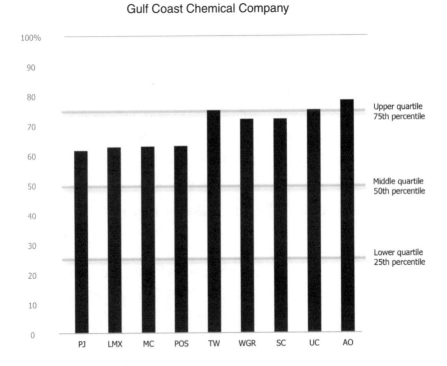

Gulf Coast Chemical Company

Organizational Dimension	Team Dimension	Safety-Specific Dimension
PJ - Procedural Justice LMX - Leader-Member Exchange MC - Management Credibility POS - Perceived Organizational Support	TW - Teamwork WGR - Workgroup Relations	SC - Safety Climate UC - Upward Communication About Safety AO - Approaching Others About Safety

Figure 10-6. OCDI scores for Gulf Coast chemical company. Scores are expressed as percentiles showing comparison to other organizations.

of consistency. Interestingly, all scales on the Organization Dimension were lower than any other scale, below the 75th percentile. This suggests that leadership capability is indeed the improvement opportunity, and that if it is improved, the other scales will also improve. This is not a bad profile, but neither is it the picture of a world-class organization, which is what it aspired to be.

Leadership needed to regain its safety leadership position, strengthen its culture, and revitalize its employee-driven safety process.

Intervention Plan

Leadership tools were targeted at the four scales of the Organization Dimension. This would also revitalize the employee-driven safety process. Specific aspects of leadership strength were assessed using the Leadership Diagnostic Instrument, and then training and coaching were tailored to individual needs. Simultaneously, supervisors and front-line employees were trained on methods to revitalize the employee-driven safety process.

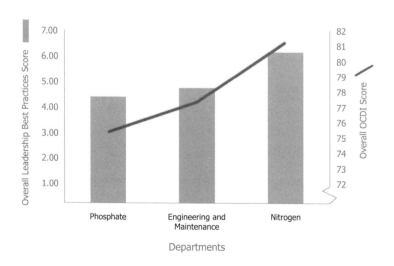

Gulf Coast Chemical Company

Figure 10-7. Overlay of leadership best practices and OCDI raw scores by leader's department at Gulf Coast chemical company. Results show a correlation between best practices and culture and climate.

This organization wanted assurance that the leadership practices measured in the Leadership Diagnostic Instrument were indeed linked directly to the scales on the OCDI. We considered this to be a reasonable concern in this case, and so we studied it. The results are shown in Figure 10-7. As expected, leadership best practices scores measured by the leadership diagnostic instrument correlate nicely with scales measured by the OCDI.

Critical Success Factors

The following factors were critical to the success of this plan:

- Selecting the right people to lead the undertaking — in particular, getting the right person to facilitate the process

- Providing ongoing leadership direction and support from the plant manager, the safety manager, and the leadership team; the willingness of the management team to commit to doing what it takes to be successful

- Support from the corporate level

Results of these initiatives will be discussed in Chapter 11.

Example 5:   Safety Culture Change in a Latin Culture: Puerto Rican Consumer Products Company

This 750-employee consumer products company in Puerto Rico had slipped to last place in divisional safety performance and was plagued by quality problems. The new plant manager realized there were significant cultural and leadership problems throughout the organization.

Objectives and Vision Statement

Diagnose and address cultural deficiencies that are impacting safety and quality. Break down cultural barriers, which include silos and traditional cultural gulfs between organizational levels. Engage all employees in a new way of relating with each other.

Results will be measured in OCDI scores, an improved level of meaningful Upward Communication, and lowered incident rates.

Assessment of Current State and Gap Definition

OCDI results are shown in Figure 10-8. The scale scores reflect a dysfunctional organization with serious cultural issues.

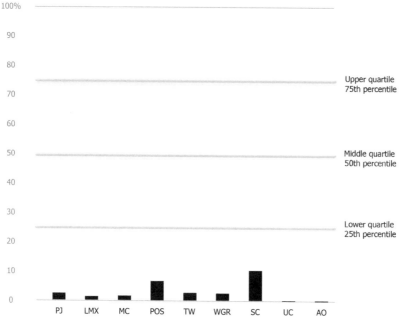

Figure 10-8. OCDI scores for consumer products company in Puerto Rico. Scores are expressed as percentiles showing comparison to other organizations.

Intervention Plan

An intensive workshop was given to senior leaders in which OCDI scores and the reasons for them were discussed in depth. A leadership development process was implemented. An employee-driven safety process was used to get employees involved.

Results from these initiatives will be discussed in detail in Chapter 11.

Summary

This chapter has focused on using diagnostic tools to understand organizational culture and safety climate in order to design appropriate intervention plans. Several examples have been given. In the next chapter we will look at examples of how intervention plans are implemented and the kind of results that follow.

• • • • •

Case Histories in Leading with Safety

- Introduction

- Shell Chemical LP

- Petro-Canada

- PotashCorp

- Consumer products company, Puerto Rico

Introduction

In the preface to this book, we referred to an outcome study, published in 1999, which tracked the results of 73 individual safety improvement projects. The data from that study are shown in Figure 11-1.

We said it was this study that started our interest in safety leadership. After finding a surprising amount of variability in individual outcomes, we did another study to determine the factors that distinguished those projects that achieved a high degree of excellence from those that were mediocre. This second study led directly to our interest in leadership and culture, and to the work described in this book.

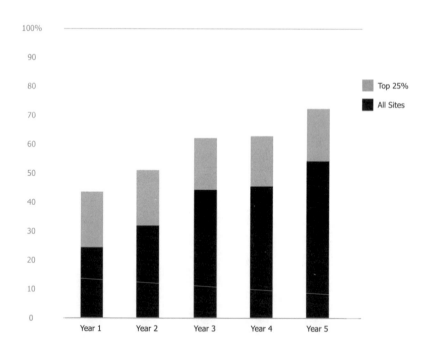

Figure 11-1. Results from the 5th edition of an outcome study demonstrating the effectiveness of BST methodology for improving safety performance. An early edition of this study has been reviewed by independent experts and published in a peer-reviewed journal. (Krause, T.R., K.J. Seymour and K.C.M. Sloat. 1999. "Long-Term Evaluation of a Behavior-Based Method for Improving Safety Performance. A Meta-Analysis of 73 Interrupted Time-Series Replications" **Safety Science.** 1999. 1-18.)

In this chapter we will review the work of four companies that have emphasized the development of safety leadership, using the tools presented in this book. We are very encouraged by the early results these companies have achieved. Figure 11-2 contrasts the average results obtained from the five-year study shown in Figure 11-1 with three projects we have enough data on to establish valid findings.

These results suggest that when safety leadership development methods are combined with employee-driven methods, the outcome is a powerful safety improvement methodology.

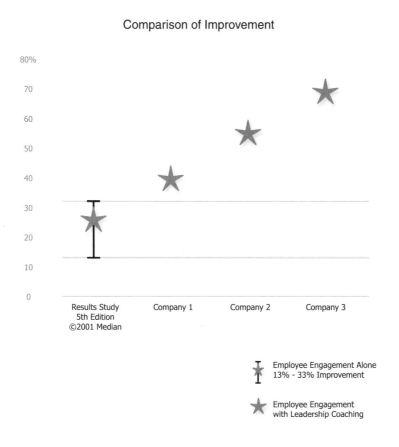

Figure 11-2. Average improvement rate of companies using employee-engagement methods alone (far left) compared to improvement rates of companies using a combination of leadership activities and employee-engagement.

Case History: Shell Chemical LP

This section was written
by Jim Spigener

In operation since 1968, the Shell Chemical plant in Geismar, Louisiana, sits on an 800-acre site on the Mississippi River. One of the world's largest producers of surfactants, the plant also produces ethylene glycol for use in antifreeze. This 530-employee facility had very good safety performance by industry standards, credited in part to a behavior-based safety process originally put in place around 1985. At that time employee-driven safety initiatives were unusual. The facility had a recordable injury rate of 6.26 and the objective of reducing it while creating an improvement process that would last over the years. Both objectives were realized. Incident rates dropped each year and the process was active and successful for more than ten years, reducing recordable injury rates to slightly above 1.0.

When a new plant manager arrived at the site in 2000, he determined that the behavior-based safety effort, while effective, needed to be re-examined. The injury rate was starting to tick up, however nothing really serious pointed directly to a coming outcome. As a result of this concern, as well as a new directive from the division leader to achieve a recordable rate of .5 by 2004, the Geismar management team began to review ways to reinvigorate the safety process, with an emphasis on integrating safety with other critical business metrics.

That site's management team came to the conclusion that the safety culture needed a renewed emphasis on the ability to recognize and mitigate risks. In addition, leadership across the Shell organization recognized the need to move the culture in all of the sites to a new level, and spearheaded this effort. The Geismar, Louisiana, site, as the first in the organization to adopt this approach, is the subject of this case history.

A Focus on Culture

The management team at Geismar wanted a world-class health, safety and environmental (HSE) system of the caliber that benefited every performance area. They recognized that this goal required an intervention that went broad and deep and that creating the desired level of functioning meant changing the culture of the whole site. Leaders needed to be able to influence the culture in a way that made safety an integral part of everyday activities. Technicians needed to be able to see safety as an organizational value and to develop and execute work practices that supported that new culture, and which enabled and motivated them to recognize and respond to exposures. Finally, supervisors and middle managers needed knowledge and tools that would enable them to develop a value for exposure recognition and respond to exposure in a way that reinforced the site's value and organizational commitment to safety.

With this new vision in mind, site leadership undertook a comprehensive evaluation of its culture as measured by the Organizational Culture Diagnostic Instrument's (OCDI) nine dimensions of organizational functioning predictive of safety outcomes (Chapter 4). The results would serve two purposes: 1) it would help the site better understand where it needed to focus attention, and 2) more specifically it would help the site's leaders identify practices they could develop to move the culture towards higher performance. Such an assessment had a third benefit: since scores on this instrument are compared to a norms database of other companies, the site would be able to see how it ranked in contrast to other organizations. It could then examine its actual organizational functioning in relation to its objective of achieving world-class performance.

The OCDI's profile was revealing. First, it showed remarkable consistency across scales, all being above the 60th percentile. The profile showed that the site enjoyed strong workgroup functioning and communication, indicating that technicians were able to work well together, and were likely to speak up among themselves about performance issues. The findings also showed that there were significant opportunities in each of the Organization Dimensions, and that the company would benefit from developing the credibility of management (Management Credibility), the employees' beliefs about the organization's support for them as individuals (Perceived Organizational Support), the level of trust between technicians and supervisors (Leader-Member Exchange), and the way technicians perceived fairness in decision-making (Procedural Justice). While the site's scores in these areas were ranked fairly high compared to other organizations, Shell Geismar felt that these scores were not high enough to support the achievement of its vision of becoming a world-class performer in HSE.

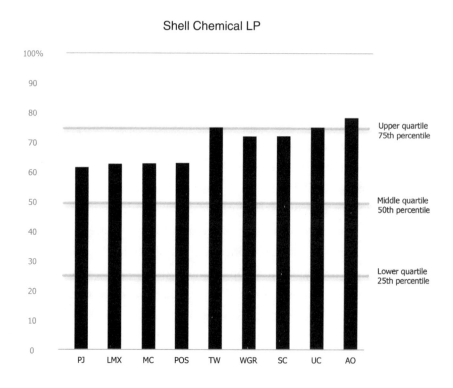

Figure 11-3. 2002 OCDI findings for Shell Chemical LP in Geismar, Louisiana. Scores are expressed as percentiles showing comparison to other organizations.

Driving the Culture from Leadership

With results of the culture diagnostic indicating improvement opportunities, Geismar's fourteen leaders agreed they needed to get aligned and to understand how they fared in execution of the leadership best practices of high-performing organizations. The group decided it would be beneficial to take 360 diagnostic assessments of each individual's leadership characteristics. The leaders used both their composite and individual scores to participate in an alignment workshop focused on helping them understand their role in improving the safety culture and defining performance targets for the group. This group focused particularly on how to develop stronger management credibility, organizational support, and communicated value for safety. Afterwards, each leader participated in a rigorous series of one-on-one coaching sessions.

As a part of this individual development, each leader was assisted in developing a personal action plan to strengthen the dimensions that provided the largest improvement opportunity for him or her. Each was coached on performing daily tasks (conducting meetings, interacting with employees) and given feedback on those interactions. Each leader ended up with a set of specific behaviors to focus on and was given regular coaching on those. Individual behaviors ranged from relationship building to open communication and collaboration with their groups.

Engaging Supervisors

The next phase of the intervention began with development activities for the site's team leaders. They played a critical role in determining day-to-day activities of the site's technicians, and would be essential in responding to exposures reported by employees. The team leaders received training in performance management skills, including recognizing barriers to safe performance, understanding systems issues that could lead to at-risk work, and working with their reports to integrate safety into the work. In addition to this work, team leaders were coached in how to respond to exposures brought to their attention, and how to shift the way they gave feedback and recognition to employees. In the new culture, employees were recognized for identifying exposures, not just for preventing an injury.

Reinventing Employee-Driven Safety

The final stage of the intervention strategy called for reviving the employee-driven safety effort. When used previously, the approach had delivered great results. However, its success and a low incident rate combined to isolate it from other production activities, and from the site's management itself. Geismar leaders recognized that the heart of the approach — engaging employees in monitoring and capturing data on risks — presented a powerful tool that would leverage the site's strong technician base. However, it would need to be adapted to move away from a focus on injury reduction toward exposure recognition and mitigation. Injury reduction relied on past incidents as a measure of likely exposure. Previous experience proved that because something hasn't happened doesn't mean it couldn't.

The revitalized employee-driven safety effort was designed to widen its scope to include attention to all exposures. Employees recruited to run the process were trained to identify exposures in whatever situations people were working regardless of whether the hazard had previously resulted in an injury. Employees were trained how to "run" scenarios. In toolbox meetings and small workgroups, employees were given a situation in which they had to identify exposures and create a solution that would reduce or eliminate that exposure. The exercises were aimed not at teaching employees specific rules, but at building fluency in the principles behind exposure identification and resolution.

Outcomes

Within six months of starting the leadership development activities and re-launching the employee-driven safety process, employees were capturing critical information about exposures. Site management now had better data with which to improve the systems and equipment employees used. In addition, leaders were already using new behaviors that supported the new culture they envisioned. At the same time, the incident rate began to drop. Within twelve months, the company's baseline incident rate had dropped to under 0.5 (Figure 11-4) achieving a rate better than required by the corporate mandate and more than a year ahead of schedule.

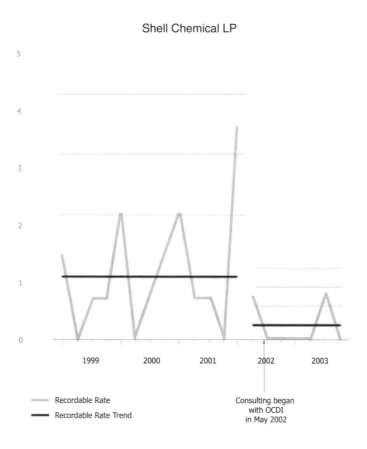

Figure 11-4. Control chart of recordable rate at Shell Chemical LP in Geismar, Louisiana.

In 2004, the site re-administered the Organizational Culture Diagnostic Instrument, (Figure 11-5). Improvements appeared in each of the nine scales, with eight of the nine scales scoring over the 90th percentile. Shell Geismar was making significant progress toward becoming a world-class HSE performer.

As a result of our work at Shell Geismar, the same style of intervention, with a tiered focus on leadership, supervisory, and front-line employees, is being adapted across the division. Each site leader is undertaking the same basic approach as the Geismar site, with significant adaptations to accommodate the unique strengths and weaknesses, structure, culture, and union affiliation of each site. (Some sites are represented and some are not.) Regardless of the

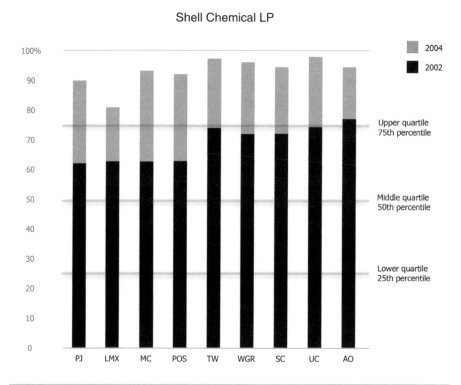

Figure 11-5. OCDI scores of Shell Chemical LP in Geismar, Louisiana, compared by year. Scores are expressed as percentiles showing comparison to other organizations.

configuration, division leaders are optimistic that the intervention strategy that succeeded so well at Geismar will allow all sites to achieve similar gains within the context of their own work situations and demands.

Case Study: Petro-Canada

This section was written in collaboration
with Rebecca Timmins

Petro-Canada is one of the largest integrated oil and gas companies in Canada. The company's five core businesses include North American Gas, East Coast Oil, Oil Sands, International, and Refining and Supply, incorporating 4,600 employees. Refining and Supply focuses on the conversion of crude oil into refined petroleum products including gasoline, diesel, asphalt, and high-quality lubricants. There is also a network of retail and wholesale outlets. It is an organization committed to profitability improvement and growth, with the goal of building shareholder value. Petro-Canada has a strong reputation for ethics, environmental responsibility, and corporate citizenship.

This case study focuses on the Refining and Supply (R&S) organization within Petro-Canada. R&S comprises three refineries — Edmonton, Alberta; Oakville, Ontario; and Montreal, Quebec — a state-of-the-art lubricants facility.

The organization recognizes that business success depends on the actions and decisions of its people and works hard to build and sustain a healthy relationship between the company and its employees based on integrity, trust, and openness. Petro-Canada's espoused values include: Results-focused, Decisive, Trustworthy, Professional, and Respectful.

The Initial Implementation: Edmonton

This case study begins in 1995 at the Edmonton Refinery where there was concern over the available tools to achieve and sustain safety excellence. Following many discussions with management and various employees, the Edmonton site began to implement an improvement process in early 1996. The objective was to continue to improve safety performance and most importantly to develop tools that would sustain a high level of performance. The implementation was a mixture of successes and challenges. Incident rates declined and the Working Interface was improved. At the same time, one of the greatest challenges was the degree of misalignment among leaders about "exactly what role managers and supervisors play in an employee-driven process."

Going Corporate-Wide

During this time a broader look at safety performance was occurring at the corporate level. Edmonton was getting attention for its work in safety and specifically for its progress with employee-driven methods, which it called Exposure-Based Safety.

After careful consideration and review of potential options, R&S decided to implement exposure-based safety at the other refineries — Oakville and Montreal, the Lubes Facility and Customer-Order Fulfillment. Concurrently, discussions were taking place at the corporate level of R&S about safety performance and the impact of leadership on safety excellence. Petro-Canada has a long history of leadership development and is proud of its efforts.

Nevertheless, there was also a sense that the tactical side of leadership — the "doing" of leaders — was not as defined or behaviorally specific as was needed to support leaders in delivering and sustaining safety success. It was agreed that to achieve aggressive safety objectives and deliver zero recordable injuries, safety leadership would be crucial. Specifically, R&S needed to develop a methodology to enable leadership at all levels to effectively support its site's exposure-based safety implementations.

The first step was to assess the existing organizational culture using the Organizational Culture Diagnostic Instrument. The next step was to align senior managers across R&S on how best to oversee and support safety and the exposure-based safety process. This included developing a common understanding among the managers and working towards agreement on the categories of key leadership practices needed to achieve the team's objectives.

Each leadership team developed a strategy to achieve success and growth. Since most leaders are skilled at defining objectives, strategies, and processes, the focus was on getting their organization to execute their strategies and processes reliably. They did this by developing behaviorally specific action plans for all managers and supervisors. To ensure successful execution of the plans, the right behaviors needed to occur at the right times and in the right ways.

Each site's focus was further defined at the individual leader level by one-to-one coaching of each leader. The focus of the coaching was to assist the leader in developing a path forward that included identifying naturally-occurring opportunities to influence safety and the exposure-based safety process. The action plans also defined behaviorally what each leader needed to do. The leaders' efforts were further reinforced through "live" observation and feedback to help share their behaviors "in the moment."

Finally, three sites (Oakville, Edmonton and Montreal) designed and implemented leadership observation and feedback processes to provide a sustainable mechanism for aligning and reinforcing leadership behavior across organizational initiatives. Carefully designed to not create a new initiative, each site developed its own inventory of leadership behaviors that cut across the many leadership development efforts that were in place, including safety leadership. Leaders learned to identify, measure and provide impactful feedback on these critical behaviors for their own and each others' benefit.

A sample of Petro-Canada site stories

Oakville Refinery

At the start of the intervention, the Oakville refinery had approximately 350 full-time and contract employees. Initial findings of the Organizational Culture Diagnostic Instrument (Figure 11-6) showed strength in relations between employees, and scores were strong in Approaching Others and Workgroup Relations. The findings showed opportunities for improvement in Perceived Organizational Support, Management Credibility, Leader-Member Exchange, and Procedural Justice, in that order. Safety Climate was also relatively low, and Upward Communication scored in the 46th percentile.

The intervention plan would include a strong employee-engagement component, but would give its greatest emphasis to leadership development, from the supervisory level all the way up to the plant manager. The site's

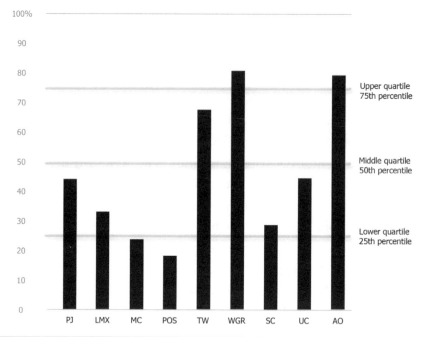

Petro-Canada
Oakville Refinery

	Organizational Dimension	Team Dimension	Safety-Specific Dimension

	PJ - Procedural Justice
	LMX - Leader-Member Exchange
	MC - Management Credibility
	POS - Perceived Organizational Support
		TW - Teamwork
		WGR - Workgroup
		      Relations
			SC - Safety Climate
			UC - Upward Communication
			      About Safety
			AO - Approaching Others About Safety

Figure 11-6. OCDI scores for Petro-Canada's refinery in Oakville, Ontario. Scores are expressed as percentiles showing comparison to other organizations.

top five managers were assessed using the Leadership Diagnostic Instrument followed by one-on-one coaching, the development of individual action plans, and leader-to-leader feedback on critical leadership behavior.

Within months of launching the intervention, the site faced a new challenge. The company announced that the small refinery would close and be transformed into a terminal with a much smaller workforce. While the exposure-based safety team had committed to continuing process activities, and the leadership team to supporting those activities, the senior leader at the site knew that successfully maintaining the course of actions as the transition from refinery to terminal occurred would require articulating a specific vision that employees could embrace. When Oakville leaders were asked, "How do we want to manage the next chapter in Oakville's history?" the overwhelming answer was: "Proud that we maintained safe, environmentally sound and reliable plant operations, proud that we treated people fairly, and proud of my personal contribution, growth, and development."

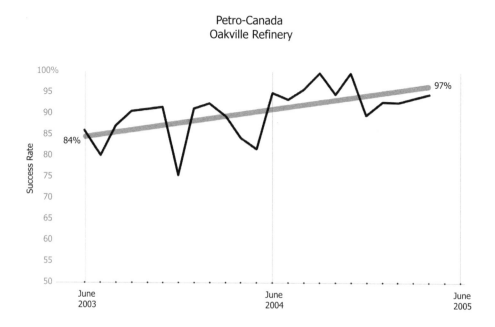

Figure 11-7. Success rate in leadership behavior samples by period, Petro-Canada's refinery in Oakville.

Between the spring of 2003 and the fall of 2004, the site logged more than 6,000 observations through its exposure-based safety process and has used that data to complete twelve action plans to remove barriers to safe work in areas including: glove use, fall protection, steam tracer burns, walking/working surfaces, respiratory protection, tools and equipment, hearing protection, and lighting. The initiative's management sponsor says these plans exemplify the teamwork and resourcefulness of the Oakville refinery.

As shown in Figure 11-8, Petro-Canada's Oakville refinery has achieved a 54% reduction in its injury rate and has been on a steady improvement trend. Senior managers at Petro-Canada are pleased with the results the Oakville

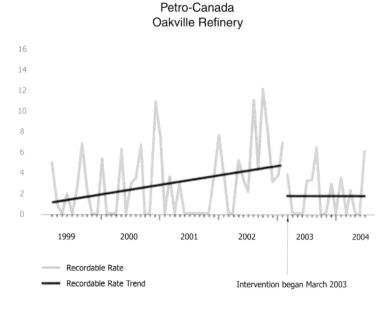

Petro-Canada
Oakville Refinery

Figure 11-8. Recordable rate at Petro-Canada's refinery in Oakville, Ontario.

refinery has achieved and are continuing to work to sustain this momentum. It aligns directly with the vision of Refining & Supply to create a solid foundation moving towards zero harm. By the spring of 2005, the Oakville site will have completed its transition from a refinery to a terminal facility. Oakville is a model for other organizations, whatever their circumstances, and illustrates the power of effective leadership and its impact on results — even in very challenging circumstances.

Edmonton Refinery

The Edmonton Refinery produces approximately 135,000 barrels of product per day. Its processes are a mixture of domestic crudes — light, sour, and synthetic — and its key product is gasoline. Three years earlier, a work stoppage had stalled the exposure-based safety effort for several months. The site had since reintroduced exposure-based safety successfully, and the effort was credited with contributing both to safe work performance and improved relations among various organizational levels and stakeholders, including the

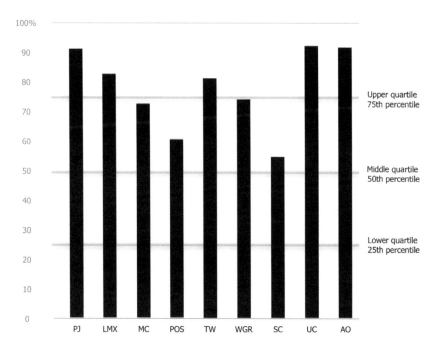

Figure 11-9. OCDI scores for Petro-Canada refinery in Edmonton, Alberta. Scores are expressed as percentiles showing comparison to other organizations.

union. Findings from the site's Organizational Culture Diagnostic Instrument show many strengths. Teamwork and Workgroup Relations are high, along with Upward Communication and Approaching Others. This suggests that the culture supports safety communications and enjoys perceptions of confidence in its leaders. At the same time, relatively low scores on Safety Climate and Perceived Organizational Support suggest the site should focus on keeping safety in the forefront.

The intervention strategy included leadership coaching, a leadership observation and feedback process, supervisor skill building, a barrier removal team to manage identified obstacles to safe work, and integration of safety activities into well-established organizational systems. Today, the Petro-Canada Edmonton refinery has achieved a recordable injury frequency of 0.55, down from 0.87 in 2000 and further improved from 2.5 in 1996. A control chart of the site's recordable rate shows a significant and sustained improvement in performance since the start of this site's interventions (Figure 11-10).

The Edmonton Refinery is a terrific example of an organization committed to long-term continuous improvement. Its story demonstrates that leadership and employee engagement can sustain safety improvement throughout periods of organizational growth and change.

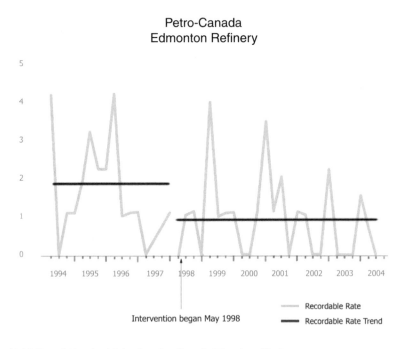

Figure 11-10. Recordable rate at Petro-Canada refinery in Edmonton, Alberta.

Case Study: PotashCorp

This section was written in collaboration
with Don Groover

PotashCorp is one of the world's premier producers of agricultural nutrients and the single largest integrated producer of potash, phosphate, and nitrogen. With the corporate office in Saskatoon, Saskatchewan, and the US office in Northbrook, Illinois, PotashCorp employs approximately 5,000 people. The organization is broken into three operating divisions, Potash, Phosphate, and Nitrogen. PotashCorp products come from seven potash locations in Canada; six phosphate operations in the United States and one in Brazil; and four nitrogen locations in the United States and one large complex in Trinidad.

PotashCorp has demonstrated that it embraces its responsibility for safety, health, and the environment. Each year the organization meets or exceeds the targets it sets out for safety improvement, and for the past ten years, PotashCorp has been steadily decreasing accident frequency. The company's long-term goals are: no harm to people, no accidents and no damage to the environment.

In 2003, PotashCorp achieved its best-ever safety performance, reaching the lowest recordable injury frequency rate per 200,000 work hours in its history. The recordable injury rate dropped 17.5%, from 2.68 in 2002 to 2.21 in 2003. This is largely the result of a company-wide focus on the key causes of accidents and the commitment of everyone who works for PotashCorp.

PotashCorp implements standardized safety programs and procedures across the company. It has developed a safety management system that outlines the basic requirements and best practices. One central theme of these programs is employee engagement, a system regarded by the company as a primary opportunity for meaningful employee involvement in safety. The employee-engagement system also provides one of several data streams that measure success at managing exposure before an injury occurs.

Leadership Development in the Potash Division

The Potash Division, based in Saskatoon, Saskatchewan, has been deploy-ing the use of leadership coaching and the concepts behind "leading with safety" to enhance safety leadership skills since 2004. In that year, three of the seven sites conducted a series of leadership workshops and put members of the leadership team through one-on-one safety leadership coaching. The remaining sites will use this approach in 2005.

The process followed at the three locations included the application of an Organizational Culture Diagnostic Instrument to get an understanding of the organizational culture and safety climate. In addition, all members of the leadership team were assessed for safety leadership practices by their direct reports. These diagnostic instruments provided the baseline information needed by the leadership team to determine the highest leverage opportunities on which to focus.

Following the collection of the baseline data, the leadership team attended a series of workshops, which focused on the concepts associated with "leading with safety." After each workshop the leaders met individually with a coach. The first session was to review the results of each leader's assessment and to discuss the organizational culture data. Then the leader and coach worked together to develop a list of leadership practices on which the individual could focus to enhance the culture and to improve the way he or she was perceived by direct reports as a safety leader and as a leader in general.

The Potash Division has seen promising results. While all but one of the seven locations in this division have used an employee-engagement process successfully over the years, in 2004 only the three sites receiving coaching showed a significant change in their incident rates — an average improvement of 40%. In comparison, those sites not receiving coaching showed no significant change in the same period.

Leadership Development in the Phosphate Division

The White Springs, Florida, operation of PotashCorp had already been named the 2003 Agri-Business of the Year when it became an early adopter of safety leadership enhancement. Made up of three major facilities over a four-mile radius, the 900-employee operation has an annual capacity of 3.6 million tons of phosphate rock and one million tons of phosphoric acid. Maintaining this level of production, and its status as the low-cost producer in the industry, is serious business that takes high-functioning leaders at all levels in three facilities.

When the White Springs operation implemented an employee-driven safety system early in 2003, it recognized the need to develop even better coordination across areas and functions as it captured data on exposure to risk. Each location had its own hourly facilitator to oversee process activities. However, managing resources across such a large area required finely-tuned alignment on what the company wanted to accomplish and how it would do it.

Cultural Assessment

The site used the Organizational Culture Diagnostic Instrument to get an assessment of its culture. The instrument, completed by all the site's employees, found that overall scores clustered around the 50th percentile, with Perceived Organizational Support and Upward Communication standing out as the areas of greatest opportunity.

Following this initial assessment, senior leaders used the Leadership Diagnostic Instrument to measure how well they demonstrated the seven best practices of leadership. The leadership team as a whole rated very high with scores across the seven scales at or above the 95th percentile. While there was significant variation within the individual results, the leadership team was viewed very positively on these seven scales by direct reports. The results suggest that the top level of leadership is sending a clear message, but the next level down isn't passing it on.

Key Leadership Practices

Using this baseline information in addition to what they learned during a day-long workshop, the leaders decided to identify a set of key leadership practices designed to address the issues. These practices would become the focal point of a series of communications targeted at creating alignment and fostering a stronger safety vision at lower levels in the organization. The leadership team knew it must demonstrate a strong level of commitment

PCS Phosphate

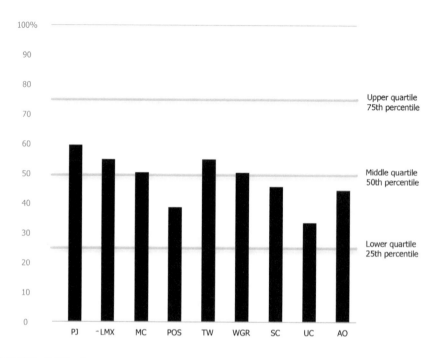

Figure 11-11. OCDI scores for PSC Phosphate in White Springs, Florida. Scores are expressed as percentiles showing comparison to other organizations.

to the fledgling employee-engagement effort which had started earlier in the year.

The leadership team spent a day together discussing the importance of safety to the employee, to the leaders, and to the middle level managers in the organization. They learned that employees view everything through the lens of culture, including any actions the company may or may not take, as well as each individual's leadership efforts. If employees had a pessimistic view of the company or its leadership team, leaders had to understand that everything they did would be seen through a "foggy" lens. Therefore, any message they sent might either not be received at all or not be received in the spirit in which it was intended.

Based on such knowledge, the leaders decided to develop a list of principles to define their actions. These principles would become a common theme for all leaders in the organization and a benchmark by which actions and decisions would be judged. The team considered close to seventy principles and selected four:

- Uphold the safety regulations even if cost or production is at stake
- Communicate frequently and effectively up, down, and across the organization
- Ensure that people have the information, authority, and resources they need
- Treat others with dignity and respect

With a clear picture of what they wanted their leadership to look like, they then worked individually with a coach to design personal strategies for improving their own interactions with those who reported to them, and for enacting the selected principles.

Outcomes

Less than a year after starting the employee-engagement effort and the development activities targeted at enhancing safety leadership skills and practices, the site saw a statistically significant reduction of 50% in OSHA incident rates. This included a six-month streak without a recordable injury. Within just a few months of identifying the key practices and starting their personal action plans, most leaders were able to document changes in their relationship with departments, showing that the new safety vision was working.

An additional challenge arose when the site experienced three serious injuries in early 2004. The site responded to these events quickly by assembling a team of site leadership, corporate leadership, and outside resources to investigate the underlying causes of the events. This team also focused on finding whether the current safety management system was capable of capturing the upstream factors to these events, and assessing the robustness of these systems. Based on the findings, the site took immediate and long-term actions that addressed identified areas of improvement. Incident rates stabilized shortly thereafter.

Overall the White Springs site has maintained a 43% improvement (Figure 11-12) in recordable rate from the time the leadership activities first began. In 2005, the site plans to build on this work by re-administering the OCDI and expanding the leadership coaching activities to the superintendent level.

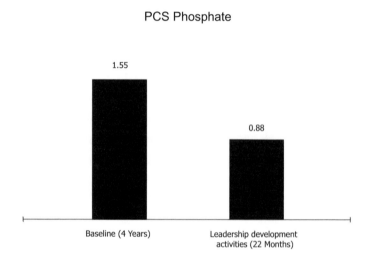

Figure 11-12. Average recordable rate from before and after intervention at PSC Phosphate in White Springs, Florida.

Case Study: Puerto Rican Consumer Products Manufacturer
This section was written in collaboration
with Kenneth Jones, Ph.D.

This consumer products manufacturer in Puerto Rico is one of eight plants in the personal care division of an international corporation, the only such site located outside the continental United States. The Puerto Rico plant traditionally manufactured two product lines: fragrances under contract to another company as well as cotton swabs. In late 2002, the site converted operations almost exclusively to the manufacture of cotton swabs and today produces all of the corporation's brand product sold in North America. Current site population is approximately 700 employees.

The Environmental Health & Safety (EHS) function is managed locally with support from the division's corporate office in the United States. While previously a responsibility assigned to the site's engineering group, EHS now stands on its own and its manager reports directly to the site manager. The production shift in 2002, in combination with staff reductions and result-ing shift changes, put the Puerto Rico site under tremendous pressure to perform. This may have impacted the incident rate. Historically, the plant was among the top performers in its division. By 2002, however, the site's accident frequency rate had quadrupled from 0.25 to 1.03. In addition to this, quality issues began to plague the site, and the site's population, which was already strongly divided along salaried/hourly lines, began to show signs that it lacked trust in management.

History of the Improvement Initiative

The new plant manager at the site recognized that the downstream safety indicators were symptomatic of a deeper problem — the plant suffered from weak leadership at the managerial and the supervisory levels. While many of the site's managerial staff were highly competent in their areas of expertise, hourly employees were underreporting incidents and even avoiding raising safety concerns out of fear of negative repercussions. The plant manager knew he needed a comprehensive approach to safety improvement that would establish a strong safety culture and foster improvement in other performance areas. Prior to the current initiative, he had enlisted the help of an outside organization and arranged for the managers to spend a weekend away at a retreat. This helped leaders begin to see themselves as a team and establish a foundation for further leadership development.

The division's headquarters expressed its support of a comprehensive effort to improve the safety performance at the Puerto Rican site. Rather than dictate a specific solution, however, the division set the expectation that the site's plant manager and his team would identify a strategic approach that met the site's unique needs. In turn, the division would provide its full support.

In order to design an effective safety intervention, the site administered the Organizational Culture Diagnostic Instrument. In addition to establishing a baseline of its current culture and safety climate, the instrument would help identify specific areas that might require greater or lesser support in the intervention strategy. A special component was added to measure whether accidents and incidents were being reported regularly.

In parallel, Leadership Diagnostic Instruments were administered to a select group of senior leaders to get a picture of the perceptions of their leadership styles. The results would serve as a launching point for working with these leaders individually.

Results of the Organizational Culture Diagnostic Instrument showed that the site suffered from low functioning along most of the variables measured. Based on these findings, a comprehensive integrated approach was designed that would address leadership and supervisory issues as well as measurement and feedback of the Working Interface.

Structure of the Improvement Initiative

The intervention was designed such that all the elements — leadership, supervisory, and employee — would be integrated and mutually supportive. Thus, work would begin with the senior leadership team at the same time as the initial training for the employee-led portion of the process. In this way there would be no doubt as to the importance of the initiative — it would be supported, and the message from the senior leadership team would be unmistakable. A cross-section of employees was chosen to participate on a

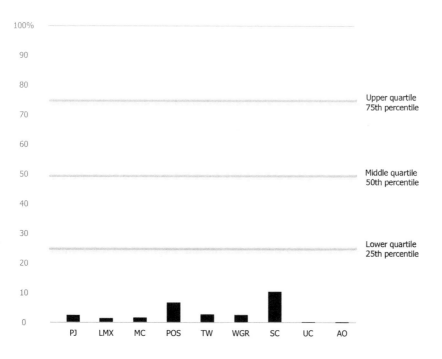

Figure 11-13: OCDI scores for consumer products manufacturer in Puerto Rico. Scores are expressed as percentiles showing comparison to other organizations.

steering team, a group that would be responsible for the implementation of the data gathering and feedback process. Supervisors would be trained shortly before data gathering began, so they could better support that portion of the initiative.

A senior leadership workshop was followed immediately with one-on-one coaching sessions. In these sessions, the participant leaders identified areas in which they wished to improve or make a change. The consultant's role was to help the leader identify specific actions to accomplish the objective. Subsequently, the leader and the consultant had occasional contact to follow up on the planned items.

A supervisory effectiveness course was implemented for all supervisors at the site, and the senior leadership team reinforced its lessons by communicating its continued expectation that attention to safety must be ever-present.

Dynamics and Challenges

Findings from the diagnostic instrument showed a sharp contrast in the scores of Organization Dimensions by level, presenting a significant challenge to improvement activities. For example, while managers saw themselves as having a high commitment to safety, hourly employees perceived that production was considered more important (Perceived Organizational Support and Safety Climate). The level of mistrust of management by hourly workers was high (Management Credibility), and workers reported they were reluctant to communicate safety concerns upward in the organization out of fear of retribution (Upward Communication). Hourly workers also suggested they would avoid reporting injuries if possible, for the same reason.

In part, this fear was justified: the atmosphere was highly punitive, as injured personnel could count on receiving disciplinary action if involved in an accident.

The plant manager recognized that these perceptions could only be changed through frequent and positive contact between managers and the hourly workforce. Following the initial coaching session, the plant manager set the expectation that all managers would develop a plan for how they would increase the instances of positive face-to-face contact with workers. At the same time, a concurrent TPM (Total Production Maintenance) initiative required managers to "adopt" a module on the production floor. Managers would have to get to know the workers on all three shifts of their adopted module, seek to understand their concerns, and offer support. This required them to come in during the night shifts, sometimes twice. As a result of both directives, the atmosphere on site shifted dramatically. Instead of little or no contact between managers and the workforce, open communication was becoming the norm.

Also important to creating a culture change would be integrating the employee-engagement initiative with other elements in the site's EHS processes. Data collected on the Working Interface was employed rigorously in removing exposure to hazards. In one instance the site undertook a long-term and expensive machine-guarding program. The initiative's management sponsor emphasized the participation of front-line workers in reducing exposure to hazards to instill a sense of ownership. He also initiated an incident investigation process in which incidents were dramatized and analyzed, and oversaw the integration of the employee-engagement effort with the company's TPM system.

Outcomes

Less than two years after undertaking this multi-level intervention, the site has reclaimed its position as the division's top safety performer. Incident rates have dropped dramatically and continue to drop (Figure 11-14). The employee-driven safety effort has been accepted as an integral part of the

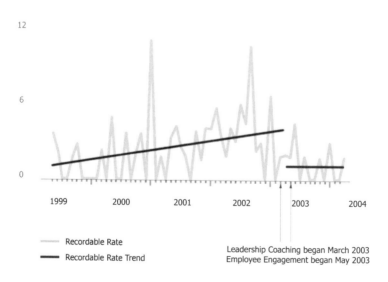

Figure 11-14. Control chart of recordable rate at consumer products manufacturer in Puerto Rico.

site's culture. Injury reporting has increased, and incidents are now being reported consistently — signs that trust of management is increasing and that managers are using new skills to strengthen the safety culture. The site-wide solution has helped salaried and hourly employees alike to unite around company objectives and make a meaningful and sustainable change in the way they work.

Lessons Learned

One of the most vital lessons of this initiative is the prime importance of a senior leader's vision and direction. Without the plant manager's strong and positive support, it would have been impossible to achieve and sustain the safety results seen at this site. Strong leadership also ensured that initiatives at the supervisory and front-line worker level were sufficiently supported with clear and consistently communicated expectations.

Another important lesson is that the employee-engagement approach is not mutually exclusive of other safety initiatives or even of other site-wide initiatives. In this case the approach was integrated both with various elements of the safety system and with TPM in such a way that the various elements became mutually supportive.

• • • • •

Chapter 12

NASA's Approach to Transforming its Organizational Culture & Safety Climate

This chapter was written
by Scott Stricoff

- Assessing the existing culture and climate

- Findings

- The intervention
    The importance of values
    Addressing culture and climate
    How leaders drive culture change

- The culture change plan

- Results
    Glenn Research Center and Stennis Space Center
    Johnson Space Center

The National Aeronautics and Space Administration (NASA) was established in 1958 to lead efforts in space exploration and aeronautics research. Today NASA has roughly 19,000 employees at its headquarters and nine Centers throughout the U.S., and more than 5,000 additional staff at the Jet Propulsion Laboratory which is operated for NASA by the California Institute of Technology. NASA's programs in space exploration, space science, and aeronautics research are widely known, with some of its most visible programs including the Space Shuttle and the International Space Station.

On February 1, 2003, the Space Shuttle Columbia and its crew of seven were lost during their return to Earth. A group of distinguished experts was appointed to comprise the Columbia Accident Investigation Board, which spent six months conducting a thorough investigation of the accident.

The Accident Investigation Board issued its report in August 2003 with findings focused on three key areas: 1) systemic safety, cultural, and organizational issues, including decision-making, risk management, and communication; 2) requirements for returning safely to flight; and 3) technical excellence. The Board found that NASA's culture and related history contributed as much to the Columbia accident as any technical failure. Specifically, the Board identified the following organizational cause of the Columbia accident:

> "The organizational causes of this accident are rooted in the Space Shuttle Program's history and culture, including the original compromises that were required to gain approval for the shuttle program, subsequent years of resource constraints, fluctuating priorities, schedule pressures, mischaracterizations of the Shuttle as operational rather than developmental, and lack of an agreed national vision. Cultural traits and organizational practices detrimental to safety were allowed to develop, including: reliance on past success as a substitute for sound engineering practices (such as testing to understand why systems were not performing in accordance with requirements/specifications); organizational barriers that prevented effective communication of critical safety information and stifled professional differences of opinion; lack of integrated management across program elements; and the evolution of an informal chain of

command and decision-making processes that operated outside the organization's rules." [1]

The Board made specific recommendations calling for a number of structural changes to the organization and identified a number of gaps in leadership practices important to safety. While there were no recommendations explicitly addressing leadership practices, the report identified many examples of gaps in the leadership practices that support safety, such as:

- Failing to follow NASA's own procedures

- Requiring people to prove the existence of a problem rather than assuming the need to assure there was not a problem

- Creating a perception that schedule pressure was a critical driver of the program

As a result of the Accident Investigation Board investigation and related activities, NASA established the objective of completely transforming its organizational and safety culture. At a minimum, it targeted making measurable progress in changing its culture within six months and having broad changes in effect across the Agency in less than three years. The six-month marker was identified as particularly critical as the Agency prepared to return to flight.

After reviewing proposals from more than forty organizations, NASA selected our firm in January 2004 to assist in the development and implementation of a plan for changing the culture and the safety climate Agency-wide. We were asked to provide for a systematic, integrated, NASA-wide approach to understanding the prior and current safety climate and culture norms, and to diagnose aspects of climate and culture that did not support the Agency's effective adoption of changes identified by the Columbia Accident Investigation Board. We were further asked to propose a course or courses of action to change behaviors and to introduce new norms that would: 1) eliminate barriers to a safety culture and mindset; 2) facilitate collaboration, integration, and alignment of the NASA workforce in support of a strong

---

[1] *Columbia Accident Investigation Board Report.* August 2003. Vol. 1, Chapter 7: 177.

safety and mission success culture; and 3) align with current initiatives already underway in the Agency.

We began with an assessment of the current status, and the development of an implementation plan. NASA asked that both be completed within thirty days. Following the assessment and the development of a plan, we began implementation. The result: significant progress towards the longer-term goal of strengthening NASA's culture. This chapter describes the assessment and its results, the plan implemented to influence the culture, and the results obtained from that plan after the initial six-month period.

Assessing the Existing Culture and Climate

Before we could change anything, we first had to understand the current culture and climate at NASA and identify focus areas for improvement. We approached this task with the belief that there was much that was positive about NASA's culture. Our challenge was to build from those positive aspects, strengthen the overall culture, and at the same time, address the issues raised in the Accident Investigation Board report.

In undertaking this work, we focused on the difference between "culture" and "climate." By culture we mean the shared values and beliefs of an organization — commonly described as "the way we do things around here." The culture can also be thought of as the shared norms for behavior in the organization, often motivated by unstated assumptions.

Climate refers to the prevailing influences on a particular area of functioning (such as safety) at a particular time. Thus, culture is more deeply embedded and long-term, takes longer to change, and influences organizational performance across many areas of functioning. Climate, on the other hand, changes more quickly, and more immediately reflects the attention of leadership.

The significance of this distinction for NASA was that in the aftermath of the Columbia tragedy there was a strong safety climate; however, we were concerned that in the absence of properly focused efforts, the culture would not change, and over time the safety climate was likely to be compromised by the inevitable schedule, budget, and operational pressures that occur in any organization.

As described below, the culture assessment was based on review of previous work, a survey of NASA employees, and a program of interviews.

Previous Studies

In late 2003, NASA Administrator Sean O'Keefe commissioned a detailed review of the Columbia Accident Investigation Board report to determine which recommendations, observations, and findings had Agency-wide applicability to NASA and to develop measures to address each one. The internal NASA team that conducted this review produced a detailed report that identified a number of concrete improvement actions and recommended assignment of these actions to various units within NASA. According to the report, the team had focused on the organizational (as opposed to physical) causes identified in the Board report, but it "did not do a broad, in-depth assessment of the cultural changes needed to address the organizational causes."

The NASA team's recommendations were divided into seven major topics:

- Leadership
- Learning
- Communication
- Processes and rules
- Technical capabilities
- Organizational structure
- Risk management

The team recognized that there was a broader need for culture change that they were not addressing. According to the report, "Some of the recommended actions are those one might expect in an organization trying to change its culture, but the goals offered by the Team are intended only as a first step in the process."

The NASA team also reviewed previous culture surveys conducted at the Agency to provide historical perspective for this assessment.

During 2003, the Federal Office of Personnel Management (OPM) conducted a survey throughout the Executive Branch entitled "Best Places to Work." This survey measured employee attitudes about various aspects of the government's agencies and resulted in an overall ranking of agencies and locations within agencies. NASA ranked highest among all agencies, and several NASA locations were on the list of the top ten locations in the entire federal government. The survey found strengths in teamwork, employee skills-mission match, and strategic management. It was also designed to identify areas in which each agency could make improvement, and at each NASA center the general category of "Leadership" was identified as an improvement target.

These findings were generally consistent with results NASA had obtained in its own previous surveys. While NASA had not conducted an Agency-wide culture survey in many years, there had been such surveys at several of the individual Centers within the last few years. These surveys identified leadership as a top area for improvement. However, they had not clearly defined the nature of the leadership improvement opportunity.

Safety Climate and Culture Survey

We conducted a specially modified version of our Organizational Culture Diagnostic Instrument (OCDI) at all 11 NASA locations. We asked all NASA employees plus Jet Propulsion Laboratory (JPL) employees to complete the survey via a web-based link. As previously described in Chapter 4, the OCDI measures the underlying organizational determinants of organizational culture and safety climate.

We administered the survey to solicit information about mission safety, which was defined as follows: "the prevention and avoidance of injury or damage to the mission or its hardware in all aspects of NASA missions."

In addition to the basic survey scales, we added questions specifically designed for use in NASA. Those questions were designed to evaluate the current situation in comparison to the desired state and to gather data on several specific culture-related issues raised by the Accident Investigation Board report.

An overall response rate of 45.2% was obtained for NASA employees, comparable to response rates obtained on previous NASA culture surveys. We evaluated potential response bias in the sample of people who responded, and these tests indicated that the respondent group was comparable to the overall NASA population.

Agency-wide response to the basic survey scales is shown in Figure 12-1 (percentile scores) and Figure 12-2 (raw scores). The percentiles in Figure 12-1 reflect comparison of NASA with a normed database compiled using this survey.

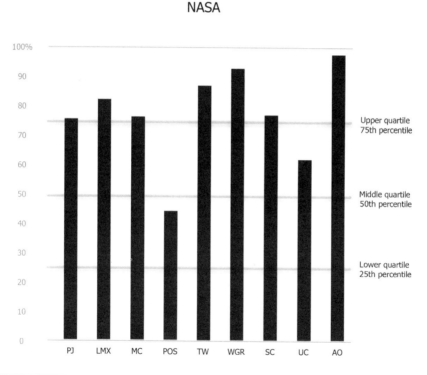

Organizational Dimension	Team Dimension	Safety-Specific Dimension
PJ — Procedural Justice	TW — Teamwork	SC — Safety Climate
LMX — Leader-Member Exchange	WGR — Workgroup Relations	UC — Upward Communication About Safety
MC — Management Credibility		AO — Approaching Others About Safety
POS — Perceived Organizational Support		

Figure 12-1. Combined OCDI scores for NASA showing overall percentiles for all locations.

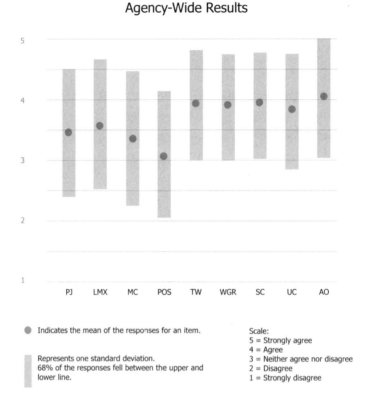

NASA
Agency-Wide Results

● Indicates the mean of the responses for an item.

Represents one standard deviation.
68% of the responses fell between the upper and
lower line.

Scale:
5 = Strongly agree
4 = Agree
3 = Neither agree nor disagree
2 = Disagree
1 = Strongly disagree

Figure 12-2. Raw scores of OCDI scales for NASA (mean and standard deviation).

At an Agency-wide level, NASA scored well in relation to other organizations in the database on most of the scales comprising the survey. It scored above the 90th percentile on Approaching Others, and Workgroup Relations, and between the 80th and 90th percentiles for Teamwork, and Leader-Member Exchange. These results indicated that across the Agency there was generally effective team functioning at the local level, with employees who have the ability and inclination to speak up to peers.

NASA scored lowest on two scales: Perceived Organizational Support (46th percentile) and Upward Communication (62nd percentile). Perceived Organizational Support (POS) measures employees' perceptions about the organization's concern for their needs and interests. Those perceptions in turn influence beliefs about the organization's values for safety. This influences employees' willingness — or unwillingness — to raise safety concerns. Upward Communication (UC) measures perceptions about the quality and quantity of upward communication about safety, the extent to which people feel encouraged to bring up safety concerns, and the level of comfort discussing safety-related issues with the supervisor.

Lower scores on POS and UC indicated areas for particular focus during the culture change effort. Senior management and the behaviors they stimulate through the management chain influence both of these dimensions. These dimensions are also a strong influence on the culture in ways that relate directly to mission safety.

Findings

To help provide context for the survey results, we conducted a series of interviews with more than 120 people at representative locations — NASA headquarters, the Glenn Research Center, and the Johnson Space Center. At each location we interviewed individual members of senior management and met with representative groups of individual contributors, and supervisors and managers. The purpose of these interviews was to provide general background to help us interpret survey data.

In general, the interviews disclosed a strong sense of dedication and commitment to the Agency's work. However, we also found frustration about a number of things.

During the interview program, we received a number of indications that there were impediments to speaking up at NASA. On more than one occasion individuals would hang back at the end of a group session and either make comments after others had left or leave written notes expressing thoughts they had not brought up in front of others. These comments tended to be on the topic of barriers to communication. This was consistent with the Upward Communication survey result and indicated that there was a group of non-managers within NASA who felt that open communication was impeded.

We also heard many comments indicating that not all managers and supervisors had the leadership skill levels that many considered appropriate.

A common theme was the issue of respect for individuals and the need for some managers to act in ways that better reflect that value.

Safety & Mission Success Week Data

In November 2003, nine months after the shuttle disaster, NASA held Safety and Mission Success Week. During this week each Center Director was asked to collect feedback from his workforce on the Columbia Accident Investigation Board report and the issues it raised.

NASA analyzed data from the centers, identifying major themes. We received the summary of this data as the assessment report was being prepared and found it was consistent with the findings of the assessment. Several of the themes and specific issues identified were important to culture change at NASA, including:

- Lack of a process for delivering upward feedback.
  This was reflected in the survey scores for Upward Communication

- Leaders do not follow words with actions. This contributes directly to lower Management Credibility

- Message of "what" delivered without the "why." This is likely to contribute to lower Management Credibility and lower Perceived Organizational Support

- Need a culture that values and promotes respect and cooperation. This relates to Perceived Organizational Support

- Need a renewed emphasis on respect for each other, and cooperation

- Minority opinions need to be embraced — create an open atmosphere in which disagreements are encouraged and new ideas/alternatives are pursued. (This was consistent with survey findings that Upward Communication was one of the weakest scales measured)

- Contractors are treated as second-class citizens. This can result in inhibiting communications, with the potential for impeding performance excellence

Conclusions

The assessment found that the NASA culture reflected a long legacy of a can-do approach to task achievement, but did not yet fully reflect the Agency's espoused values of Safety, The NASA Family, Excellence, and Integrity. The culture reflected an organization in transition, with many ongoing initiatives and lack of a clear sense at working levels of "how it all fits together."

Examining NASA's espoused values, we found that:

Safety was something to which NASA personnel were strongly committed in concept, but NASA had not yet created a culture that was fully supportive of safety. Open communication was not yet the norm, and people did not feel fully comfortable raising safety concerns with management.

The NASA Family value was inconsistent with the fact that people felt disrespected and unappreciated by the organization. As a result, the strong commitment people felt to their technical work did not transfer to a strong commitment to the organization. People in support functions frequently did not fully understand or appreciate their connection to the Agency's mission, and people in technical positions did not fully value the contribution of support functions to their success.

Excellence was a treasured valued when it came to technical work, but was not seen by many NASA personnel as an imperative for other aspects of the organization's functioning (such as management skills, supporting administrative functions, and creating an environment that encourages excellence in communications).

Integrity was generally understood and manifested in people's work. However, there appeared to be pockets in the organization in which the management chain had sent signals — possibly unintentionally — that raising negative issues was unwelcome. This was inconsistent with an organization that truly values integrity.

In summary, we identified an opportunity and needed to strengthen the culture's integrity by helping NASA become an organization that lives the values.

The Intervention

Overview

Based on this assessment, we recommended that the culture change initiative should build on the strengths shown in the safety climate and culture survey. NASA employees generally worked well as teams, liked and respected each other, and felt comfortable talking to peers. These strengths could be harnessed to create reinforcement mechanisms for behaviors that support the Agency's values and desired culture.

In addition, we recommended that the culture change initiative should focus on helping managers and supervisors maintain an effective balance between task orientation and relationship orientation. At NASA many managers had a natural inclination toward task orientation, which is not unusual for technical organizations. However, strong task orientation at the expense of relationship orientation can lead to inhibition of Upward Communication and weak Perceived Organizational Support. By taking steps to help managers and supervisors improve their balance between task and relationship orientation, NASA could move toward integrating its values of Safety and People and create a culture that would more effectively support the Agency's mission.

We believed that NASA needed to avoid falling into the organizational "trap" of viewing its response to the Board report purely in a project-driven manner. The NASA culture tended to think in terms of identifying problems and solving them through discrete projects. Over the years NASA had proven to be outstanding at defining and executing projects. However, a project is, by its very nature, something that has a start and an end. If it came up with separate projects to address specific issues in the report, the Agency could fail to address the underlying culture issues that gave rise to many of the problems in the first place. This may explain why safety climate changes observed after previous accidents (e.g., the Shuttle Challenger accident) did not generalize and become part of the ongoing culture.

To address NASA's needs and build on its strengths, we developed a culture change plan based on one core concept: *Organizational values must underlie the definition of desired culture.*

The Importance of Values

Values underpin everything an organization does to ensure that objectives are reached. They help inform everyone in the organization about the considerations that should be reflected in day-to-day actions and decisions. Values set out the basis for the strategic considerations necessary for success and help ensure that everyone understands the organization's expectations of them.

An organization cannot create specific rules covering every situation and variation. In the complex world in which NASA functions, the Agency must be able to rely on individuals making independent judgments about unexpected and unforeseen situations. Having organizational values that are well understood and embraced by everyone will reduce the variability with which these judgments are made.

According to the assessment results, there was no uniformity of adherence to the espoused organizational values that would lead to safety performance excellence. The implementation plan recognized the importance of values for a safety-supporting culture being widely disseminated and embraced within NASA and actively reflected in the leadership practices of individuals at all levels of the organization.

Addressing Culture and Climate

Both climate and culture are important. While identifying values was an important first step, building these values into the fabric of the Agency required transforming the culture.

Organizational climate often changes very quickly after a significant incident, but the underlying organizational culture may not change sufficiently to prevent further incidents. Since climate that is inconsistent with culture will not be sustained, a favorable safety climate following an incident does not assure real improvement unless steps are taken to shift the culture.

As we developed the implementation plan, the current climate for safety in NASA was very strong and favorable. Since favorable organizational climate is a condition for successful culture change, this situation presented a limited-time opportunity to introduce new principles that could lead an Agency-wide cultural change initiative.

How Leaders Drive Culture Change

The key to changing culture is through leadership. Leaders influence safety through what they do and what they don't do. They can express this influence intentionally or unintentionally. However, leaders with the right knowledge and skills can move the culture in desired ways and do so with accelerated results. Therefore, the key is to make leaders more effective, and the best way to do that is through the use of behavioral tools.

Using Behavioral Tools. Behavioral tools are the most practical and effective way to transform culture; culture changes when new behavioral norms are established. Because behavior is definable and measurable, it lends itself to change efforts. By using behavior-based tools, organizations can undertake very concrete and specific initiatives to accelerate cultural transformation and can measure progress toward results.

Behavioral tools may be used to create accelerated change within organizations as well as to ensure that future leaders are selected and developed to sustain the desired culture. Our assessment results confirmed the opportunities to use these tools for the change desired by NASA.

Focusing Culture-Change Efforts. There should be one, single culture change initiative. NASA was in a period of change, with many active teams and task forces. Many of these had identified issues that relate to culture, and this raised the possibility that there could be overlapping, or even contradictory initiatives.

For culture change at NASA to be successful, there needed to be a consistent culture change initiative that incorporated all of its culture-related issues.

The Culture Change Plan

The specific plan we developed for the initial six-month period was designed to begin the culture change while validating the adaptation of the approach to fit NASA. To do this we focused on three NASA locations — the Glenn Research Center, the Stennis Space Center, and two large directorates of the Johnson Space Center (Engineering and Mission Operations). These organizations collectively comprised approximately 3,600 people.

Changing the culture involves two thrusts. The first engages leadership and individual contributors in changing the current cultural environment; the second assures that the culture is sustained by grooming future leaders who can support the desired culture. This initial phase of the effort focused on the former objective.

At the outset, NASA's senior leadership re-examined the organization's core values and reaffirmed those to which the Agency aspires. Those values were used to articulate a vision of the future state that would exist following successful culture change:

> "The objective of this effort is to strengthen the organizational culture and safety climate at NASA. In this desired future state, each individual feels highly valued as an individual and knows that his or her contributions are appreciated. Everyone at the Agency, in all roles and at all levels, understands the important ways they contribute to the Agency's exciting mission, feels like an integral part of the larger Agency team, understands the way that others contribute to the larger team effort, and is committed to the success of the Agency and its overall mission. Managers and executives at every level of the Agency, from top to bottom, routinely treat people with respect. People are comfortable in raising issues, and confident that the issues raised are considered and appropriately factored into decisions. There is a high level of trust in management, and a sense that management, in turn, trusts each individual.
>
> In this desired future state, safety is widely recognized as an integral component of mission success, and is considered by every individual in everything they do. The Agency is recognized for its pursuit and outstanding achievement of cutting edge endeavors, as well as its extraordinary safety record, all of which are understood as compatible goals."

In designing a strategy to achieve the culture change objective, we began with the recognition that culture is a reflection of shared perceptions, and beliefs and behaviors. It is related to unstated assumptions. If we change those perceptions and beliefs, we change culture.

Individuals' perceptions and beliefs are influenced by a variety of factors subject to intervention. For example, perceptions and beliefs about the organization are strongly influenced by individuals' interactions with their immediate supervisors. These interactions inform the individual about the organization's real values and shape his or her views about the organization. There are dozens of these interactions each week. A change in the leadership behavior of the immediate supervisor will influence culture, but is unlikely to occur unless there are changes in the leadership behavior of that supervisor's supervisor. Similarly, we must change behavior up through the leadership chain.

To change individuals' perceptions and beliefs, we wanted to change their supervisors' leadership behaviors to more consistently reflect behavior that reflects the desired culture. The new behaviors we wanted to encourage in NASA's first-line supervisors — Branch Chiefs — were a set of critical behaviors that exemplify NASA's core values. The behaviors we wanted to encourage up through the chain of command — through Division Chiefs, Directors, and Center Directors — were those that exemplify the values and encourage the use of these behaviors by subordinate managers.

There is a large set of behaviors that supports NASA values, including both leadership behaviors and individual contributor behaviors. To change culture we needed to focus on a manageable subset of those behaviors, selected for their leverage in affecting perceptions and beliefs related to areas in which we wanted the culture to change. For example, survey results showed that NASA's culture was strong in the area of Workgroup Relations. While there are behaviors related to Workgroup Relations, those were not the ones on which we chose to focus as they were already comparatively strong. However, in an area like Upward Communication, where NASA needed to improve, the related leadership behaviors would be considered "critical behaviors." Critical behaviors for NASA at this time related to communication, consideration for individuals, management consistency (credibility), and decision-making.

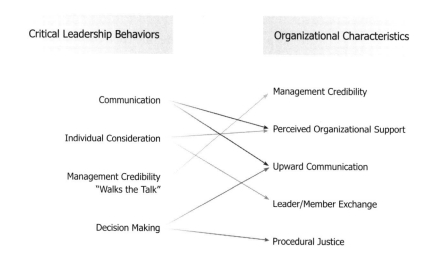

Figure 12-3. The relationship between critical leadership behaviors and key organizational characteristics.

Critical behaviors were identified based on a variety of data sources such as the Columbia Accident Investigation Board report, the OCDI, NASA's internal review of the broad applicability of the Board recommendations, and Safety & Mission Success Week findings. A foundational set of critical leadership behaviors was identified based on those data sources. This foundational set of critical behaviors was then reviewed by each location at which the culture change effort was to be implemented. This review verified the relevance of the behaviors to each location and developed examples of how each behavior was manifested at the location, to embellish the definition for local use.

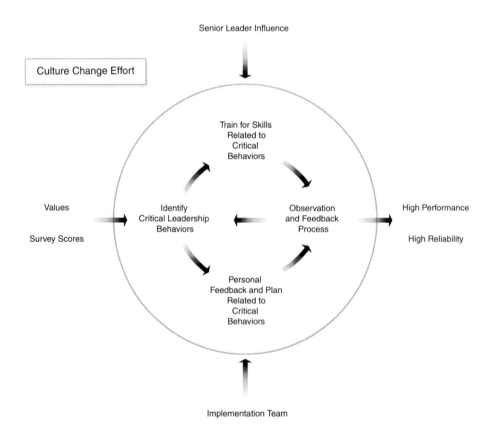

Figure 12-4. Implementation strategy for individual NASA locations.

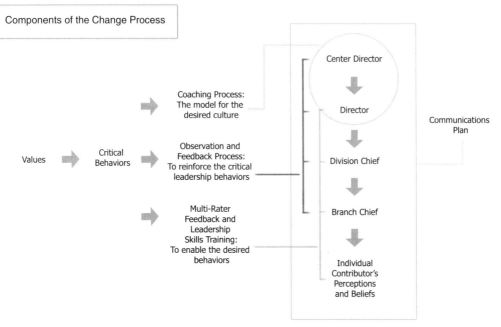

Figure 12-5. Key components of the change process at NASA centers.

We designed a multi-pronged approach of specific activities that included introducing leadership coaching for senior-level leaders, implementing a behavioral observation and feedback process for all leaders, and providing multi-rater feedback and skills training for all leaders. A communications effort was also launched at each location to inform people about the changes occurring.

Coaching

The senior-most leaders in the organization have an important, but indirect, influence on the perceptions and beliefs of most individual contributors. Therefore, the senior-most leaders must possess strong leadership skills and a solid understanding of how they can exert influence. It is important that they set the direction for the culture through everything they do and that they create consequences that cause their reports to do the same. To help senior-most leaders support the culture change, we employed a leadership coaching process. This helped the leaders improve their ability to support the critical behaviors (as well as practice these behaviors themselves) and helped them learn how to meaningfully support the other elements of the change process.

The coaching process was designed to help senior leaders understand their leadership strengths and weaknesses and to work with them in developing individual action plans. The process began with a detailed individual assessment including a 360 diagnostic survey plus a series of assessment interviews with subordinates, peers, and managers. The assessment resulted in a detailed feedback report that assessed the individual's leadership style and practices. Because this report was based on information from individuals familiar with the leader and provided detailed examples of his or her leadership behavior, it filled a vacuum that most senior leaders have — a lack of direct feedback on their leadership.

The coach reviewed the feedback report with the leader and then helped to develop a coaching action plan. This plan identified areas for the leader to concentrate on, drawing on the critical behaviors, the actions needed to drive support for NASA's values, and leadership best practices. Once the plan was developed, the coach provided the leader with guidance as the coaching action plan was implemented.

The coaching process was used for senior leaders, beginning at the top of the Agency and extending down through the management chain to the senior-most levels of the Center.

Behavioral Observation and Feedback

All leaders in the organization were required to adopt and consistently use the critical leadership behaviors. A behavioral observation and feedback process was implemented to promote use of these behaviors. Leaders receiving regular, structured reinforcing feedback on their use of critical behaviors and guidance feedback on missed opportunities to use these behaviors would change their behavior. When their use of critical behaviors was encouraged by those senior to them in the organization (as a result of the coaching process), this change would be further encouraged.

Anonymous data was gathered during these observations, allowing the local implementation team to track progress in promoting critical behaviors, analyze the reasons for non-performance, and design corrective action as appropriate.

Multi-Rater Feedback

We provided each leader with individual multi-rater survey feedback to help him understand which types of behavior represented existing strengths, and which represented areas for focusing improvement efforts. We used a 360 diagnostic survey to gather feedback on each individual leader's use of leadership and management best practices. Leaders attended a workshop to review and discuss the results and to develop individual action plans focused on increasing their use of leadership behaviors that supported the organization's values.

Skills Training

The objective of the skills training was to improve skills leaders need to perform the critical behaviors and support the desired culture. Managers received two days of training, which covered cognitive bias awareness and feedback skills (day 1) and influential leadership skills such as building trust, valuing minority opinion, and influencing skills (day 2). Each of these segments was explicitly tied to critical behaviors being addressed in the culture change initiative.

Communications

The fifth element of the near-term culture change process was communications, and there were two aspects of this challenge.

At the individual Centers where culture change activities were occurring, it was important that there be communication about these efforts. "What" was occurring and "why" had to be communicated at the outset. Then, as implementation proceeded, it was especially important to communicate about early indications of progress.

The specific mechanisms for this communication varied from Center to Center based on the communications vehicles available locally. Existing communications channels such as site newsletters, intranets, and all-hands meetings were used to help relay information about this effort. In addition, managers were encouraged to speak about it at their staff meetings.

More globally, it was important that NASA's overall communications reflect consistency with the culture change effort and the desired culture. Even on topics not directly related to the culture change effort, senior leaders indirectly send messages about how seriously they take the desired culture. When members of NASA's senior-most leadership spoke or sent written messages, the content of those messages needed to reflect specific consideration for the cultural undertones of the communication.

Results

For five months beginning in mid-April 2004, we worked with the Glenn Research Center, Stennis Space Center, and the Engineering and Mission Operations Directorates of the Johnson Space Center. This initial phase of work was designed to provide a mechanism to learn how best to deploy the culture change approach while meeting the objective of achieving measurable progress in six months.

As the work progressed, various forms of results data became available.

Anecdotal Data

Soon after implementation work began, we started hearing anecdotal evidence that the effort was having an effect. Examples of the anecdotal evidence are listed in Table 12-1. This evidence provided early indications that the culture change effort was beginning to have an impact.

Behavioral Data

As data began to accumulate from the behavioral observation and feedback process, we started seeing improvement in the percentage of times an observed behavior was observed being done, rather than observed as a missed opportunity. Figure 12-6 shows early data from one location. Several of the specific behaviors are showing an improvement trend. Other behaviors did

Early Anecdotal Data

"Helps me be less judgmental & see myself as others do"
    – an observer

"I wasn't sure of this thing in the beginning. Now I am convinced that it will help us; we need to support it. I have invited observers to my meetings; I encourage you to do the same."
    – Division Chief

Division Chief asks that two meetings be observed

"I found myself conducting my Branch meetings and day-to-day interactions differently as part of this effort. I am convinced that others will also change their habits; even if they are not bad right now but improvement is good."

One Implementation team had a well-known skeptic as a member. After observer training he got up and told the group that he hadn't been in favor of this, but now that he understood it he thought it was going to make a big difference.

Individuals requesting to have 360 leadership survey done to provide them with feedback

Training evaluations consistently indicating that participants arrived as skeptics and left as believers ("prisoner" to "advocate")

Division Chiefs giving each other feedback in a staff meeting, referring to the coached behaviors

Observer invited to observe MMT meeting

Table 12-1. Examples of early success indicators in the change process at NASA.

not show improvement this rapidly, but the data produced by the process provided a mechanism to know where to place emphasis in seeking further improvement.

Culture Survey

Approximately six months after the start of the culture change efforts, we administered the OCDI again to the groups where culture change work had been undertaken. This was the same survey used in the initial assessment phase of the effort, and we used the same email-prompted, web-based survey administration method.

The response rate was quite good, and at most locations it exceeded the rate obtained in the original (February) survey administration, as well as

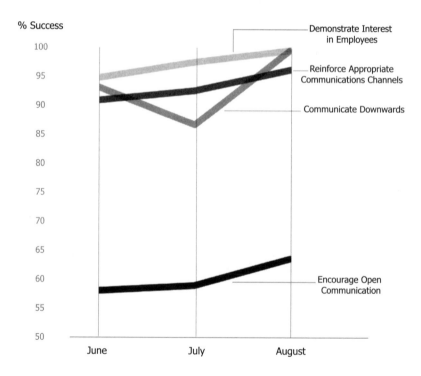

Figure 12-6. Early data from one NASA location showing improvement.

the rates obtained on previous NASA culture surveys. The response rate by location is shown in Table 12-2.

Tests to evaluate potential response bias in the sample of people who responded indicated that the sample was representative of the total surveyed population.

The Glenn Research Center and Stennis Space Center had survey scores during the initial assessment that were low compared to the NASA overall averages. The Johnson Space Center had scores that were high relative to the NASA average. The results of the intervention at these centers are interesting to compare.

Glenn Research Center and Stennis Space Center Results

All scales on the basic Safety Climate and Culture survey showed improvement at the Glenn Research Center (GRC). These results are shown in Figure 12-7 (percentile scores) and Figure 12-8 (raw scores.) The September results (after intervention) show significant improvement over the February results (pre-intervention).

	February Response Rate (%)	September Response Rate (%)
Glenn	32.4	65.2
Johnson (Engineering & MOD)	52.6	45.8
Stennis	45.2	71.5
Overall	**45.2** (Nasa-Wide)	**57.9**

Table 12-2. NASA survey response rate by month.

Figure 12-8 shows the comparison of these results with their confidence intervals. Where confidence intervals do not overlap, the differences are statistically significant.

Comparing managers' responses to non-managers' responses at GRC, we found a greater change in survey scale results among managers than among non-managers. This is consistent with what we would expect after just six months: the culture change strategy was to work with leadership as the mechanism for driving culture change. Initial activity in the culture change

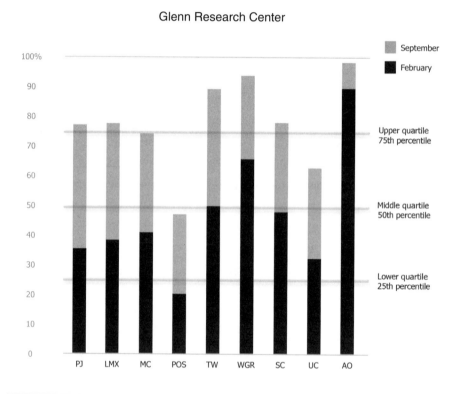

Figure 12-7. OCDI percentile scores for NASA's Glenn Research Center showing results from before and after start of intervention.

effort focused primarily on managers at all levels. After just six months, one would expect to find managers seeing greater change than individual contributors, and that is what the results indicated.

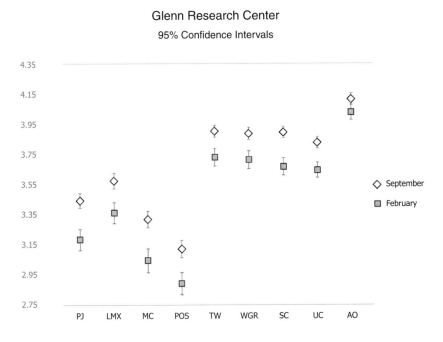

Organizational Dimension	Team Dimension	Safety-Specific Dimension

PJ   - Procedural Justice
LMX - Leader-Member Exchange
MC - Management Credibility
POS - Perceived Organizational Support
	TW - Teamwork
	WGR - Workgroup
	      Relations
		SC - Safety Climate
		UC - Upward Communication
		      About Safety
		AO - Approaching Others About Safety

Figure 12-8. OCDI raw scores for NASA's Glenn Research Center.

Scale : 5 = Strongly agree
         4 = Agree
         3 = Neither agree nor disagree
         2 = Disagree
         1 = Strongly disagree

The final question in the survey was open-ended: "What changes have you seen in NASA's culture in the last six months?" Among GRC managers, 46% of respondents provided comments, and among non-managers 44% provided comments.

In analyzing the comments provided by managers, we found that 32% mentioned specific indicators of culture improvement such as seeking input from others[2], while 10% indicated they had seen no change, and 4% indicated that the culture had worsened. Among managers providing comments, 21% indicated an improved safety climate, while 4% indicated the safety climate was worse.

Among non-managers, 22% mentioned specific indicators of culture improvement, with 16% indicating no change, and 4% indicating a worsening of the culture.

In addition to the basic survey scales, this survey included a series of NASA-specific questions. They were grouped into several thematic areas such as guiding principles for safety excellence, consistency between words and actions, cooperation and collaboration, potential inhibitors, communication, and employee connection to mission safety. All NASA-specific questions showed improvement compared to the first survey.

Results from the Stennis Space Center were very similar to those from GRC. All survey scores improved, and comments were consistent with these results.

Johnson Space Center Results

The survey was administered at Johnson Space Center (JSC) to the Engineering Directorate and the Mission Operations Directorate (MOD). The culture change efforts had been focused on these two groups during the initial phase of the process.

All scales on the basic Safety Climate and Culture survey showed improvement for these two JSC organizational units. These results are shown in Figure 12-9 (percentile scores) and Figure 12-10 (raw scores.) The September results

---

[2] Only comments mentioning changes to cultural characteristics were counted. Many other comments mentioned activities undertaken during the last six months, such as training or meetings, but descriptors of activities - as opposed to characteristics of culture - were not counted for analysis.

show significant improvement over the February results.

Figure 12-10 shows the comparison of these results with their confidence intervals. Where confidence intervals do not overlap, the differences are statistically significant.

JSC had generally high scores on most scales prior to the culture change efforts, with most scales above the 80th percentile. In the survey conducted after the initial culture change efforts, every scale showed some level of improvement. Percentile scores were high, although raw scores still showed room for improvement.

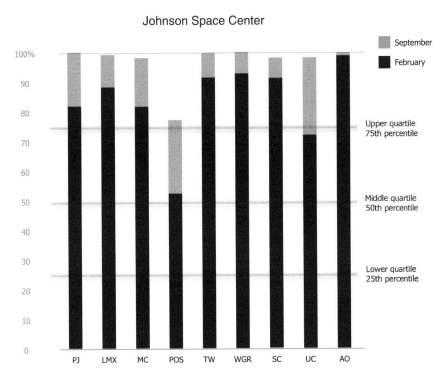

Organizational Dimension	Team Dimension	Safety-Specific Dimension
PJ - Procedural Justice LMX - Leader-Member Exchange MC - Management Credibility POS - Perceived Organizational Support	TW - Teamwork WGR - Workgroup Relations	SC - Safety Climate UC - Upward Communication About Safety AO - Approaching Others About Safety

Figure 12-9. OCDI percentile scores for NASA's Johnson Space Center showing results from before and after start of intervention.

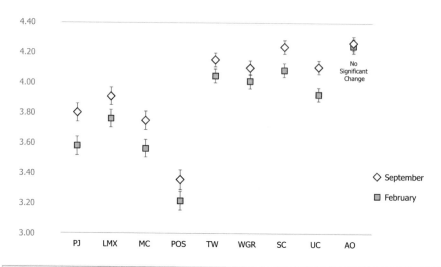

Figure 12-10. OCDI raw scores for NASA's Johnson Space Center.

Scale :   5 = Strongly agree
          4 = Agree
          3 = Neither agree nor disagree
          2 = Disagree
          1 = Strongly disagree

Comparing managers' responses to non-managers' responses, we again found a greater change in survey scale results among managers than among non-managers. As noted in the discussion of GRC results, this was consistent with what we would expect.

The final question in the survey was open-ended: "What changes have you seen in NASA's culture in the last six months?" Among JSC managers, 52% of respondents provided comments, and among non-managers, 45% provided comments.

Among the responses provided by managers, 52% mentioned specific indicators of culture improvement such as seeking input from others[3], while 7% indicated that they had seen no change, and 4% indicated that the culture had worsened.

Among non-managers, 22% mentioned specific indicators of culture improvement, with 22% indicating no change, and 3% indicating a worsening of culture. In addition, 13% indicated improvement in safety climate.

In addition to the basic survey scales, this survey included a series of NASA-specific questions. All NASA-specific questions showed improvement since the February survey.

Summary

By focusing on leadership using behavior-based tools, NASA has made a strong start in its effort to change its culture. Both survey scale scores and comments indicate that the change effort at NASA has made good progress in a brief time, but that more work remains to be done. As would be expected in the early stages of a major change effort, there appears to be a segment of the population that is seeing positive change and is optimistic about the direction the organization is moving, and another segment that is skeptical and not yet seeing what its members articulate as change. However, the overall perceptions, measured by the survey scores, indicate that there is solid movement in the desired direction.

The approach taken has built ownership for the culture-change effort among the leaders of the target groups and has produced a rapid start to the longer-term job of changing the culture. Leaders have been given new tools to help them carry the change forward, and as the effort is now being expanded to the rest of the organization, NASA is on a trajectory toward an enhanced organizational culture.

• • • • •

---

[3] Only comments mentioning changes to cultural characteristics were counted. Many other comments mentioned activities undertaken during the last six months, such as training or meetings, but descriptors of activities - as opposed to characteristics of culture - were not counted for analysis.

Bibliography

The articles and books in this bibliography discuss various aspects of leadership and safety performance. The author provides this list for readers who want more in-depth information on these subjects.

Antonakis, J., A.T. Cianciolo, and R.J. Steinberg (Editors). 2004. *The Nature of Leadership*. Thousand Oaks: Sage Publications.

Avolio, B. 1999. *Full Leadership Development: Building the Vital Forces in Organization*. Thousand Oaks: Sage Publications.

Barling, J., C. Loughlin, and E. K. Kelloway. 2002. "Development and Test of a Model Linking Safety-Specific Transformational Leadership and Occupational Safety." *Journal of Applied Psychology*. 87: 488-496.

Bass, B.M. Winter 1990. "From Transactional to Transformational Leadership: Learning to Share the Vision". *Organizational Dynamics*. 18: 19-31.

Blair, E. August 1999. "Behavior-Based Safety: Myths, Magic & Reality." *Professional Safety*.

Blake, R.R., and J.S. Mouton. 1964. *The Managerial Grid*. Houston: Gulf Publishing Company.

Den Hartog, D.N, J.J. Van Muijen, and P.L. Koopman. 1997. "Transactional Versus Transformational Leadership: An analysis of the MLQ." *Journal of Occupational and Organizational Psychology*. Leicester. 70: 19-34.

Erickson, J.A. May 1997. "The Relationship Between Corporate Culture and Safety Performance." *Professional Safety*. 29-33.

Evans, B. 1990. *Bias in Human Reasoning: Causes and Consequences (Essays in Cognitive Psychology)*. London: Psychology Press.

Fairhurst, G.T., L.E. Rogers, and R.A. Sarr. 1987. "Manager-Subordinate Control Patterns and Judgments About the Relationship." *Communication Yearbook*, 10: 395-415.

Greenberg, J. and R. Cropanzano (Editors). 2001. *Advances in Organizational Justice*. Stanford: Stanford University Press.

Hall, J. 1988. *Models for Management, The Structure of Competence, Second Edition*. The Woodlands: Woodstead Press.

Hammond, J.S., R.L. Keeney, and H. Raiffa. 1998. "The Hidden Traps in Decision Making." *Harvard Business Review*. 76: 47-58.

Hayes, B.E., J. Perander, T.Smecko, and J. Trask. 1998. "Measuring Perceptions of Workplace Safety: Development and Validation of the Work Safety Scale." *Journal of Safety Research*. 29: 145-161.

Hidley, J. H. July 1998. "Critical Success Factors for Behavior-Based Safety." *Professional Safety*. 30-34.

Hofmann, D.A. and F.P. Morgeson. 1999. "Safety-Related Behavior as a Social Exchange: The Role of Perceived Organizational Support and Leader-Member Exchange." *Journal of Applied Psychology*. 84: 286.

Hofmann, D.A. and A. Stetzer. 1998. "The Role of Safety Climate and Communication in Accident Interpretation: Implications for Learning from Negative Events." *Academy of Management Journal.* 41: 644-657.

Hogan, R., G.J. Curphy, and J. Hogan. 1994. "What We Know About Leadership: Effectiveness of Personality." *American Psychologist.* 49: 493-879.

Hurtz, G.M. and J.J. Donovan. 2002. "Personality and Job Performance: The Big Five Revisited." *Journal of Applied Psychology.* 85: 869-879.

Huselid, M.A. 1995. "The Impact of Human Resource Management Practices on Turnover, Productivity, and Corporate Financial Performance." *Academy of Management Journal.* 38: 635-672.

Judge, T.A., J.E. Bono, R. Ilies, and M.W. Gerhardt. 2002. "Personality and Leadership: A Qualitative and Quantitative Review." *Journal of Applied Psychology.* 87: 765-780.

Judge, T.A., C.A. Higgens, C.J. Thoreson, and M.R. Barrick. 1999. "The Big Five Personality Traits, General Mental Ability, and Career Success Across the Life Span." *Personnel Psychology.* 52: 621-652.

Kahneman, D., P. Slovic and A. Tversky. 1982. *Judgment Under Uncertainty: Heuristics and Biases.* Cambridge: Cambridge University Press.

Kazden, A.E. 2000. *Behavioral Modification in Applied Settings, 6th Edition.* Belmont: Wadsworth Publishing.

Kotter, J.P. 1986. *Leading Change.* Boston: Harvard Business School Press.

Kouzes, J. and B. Posner. 1995. *The Leadership Challenge.* San Francisco: Jossey-Bass, Inc.

Koys, D. J. 2001. "The Effects of Employee Satisfaction, Organizational Citizenship Behavior, and Turnover on Organizational Effectiveness: A Unit-Level, Longitudinal Study." *Personnel Psychology*. 54: 101-114.

Kraines, G.A., M.D. 2001. *Accountability Leadership: How to Strengthen Productivity through Sound Managerial Leadership*. Franklin Lakes: Career Press.

Krause, T.R., and T. Weekley. November 2005. *"A Four-Factor Model for Safety Leadership." Professional Safety*.

Krause, T.R. and J.H. Hidley. January 2005. "Feedback and Recognition." *Perspectives in Behavioral Performance Improvement*.

Krause, T.R., and J.H. Hidley. August 2004. "The Art of Collaboration." *Perspectives in Behavioral Performance Improvement*.

Krause, T.R. June 2004. "Influencing the Behavior of Senior Leadership: What Makes a Great Safety Leader?" *Professional Safety*.

Krause, T.R., and J.H. Hidley. April 2004. "The Credible Safety Leader." *Perspectives in Behavioral Performance Improvement*.

Krause, T.R., and J.H. Hidley. February 2004. "The Challenge of Vision" *Perspectives in Behavioral Performance Improvement*.

Krause, T.R. and J.H. Hidley. April 2003. "Senior Leaders in Safety: Real-World Examples of Leaders Making a Difference." *Industrial Safety & Hygiene News*.

Krause, T.R. and J.H. Hidley. March 2003. "Senior Leaders in Safety: How to Shape a Safety Culture." *Industrial Safety & Hygiene News*.

Krause, T.R. and J.H. Hidley. February 2003. "The Role of Leaders in Safety Performance." *Industrial Safety & Hygiene News.*

Krause, T.R. 2002. "Organizational Factors that Predict Safety Success." *Industrial Safety & Hygiene News.*

Krause, T.R. September 2001. "Improving the Working Interface." *Occupational Hazards.*

Krause, T.R. May 2001. "Moving to the Second Generation in Behavior-Based Safety." *Professional Safety.*

Krause, T.R. March 2000. "Motivating Employees for Safety Success." *Professional Safety.* 45: 22-26.

Krause, T.R., K.C.M. Sloat, and R.S. Stricoff. March 2000. "Leading for Safety." *Pima's North American Papermaker.*

Krause, T.R., K.J. Seymour, and K.C.M. Sloat. 1999. "Long-Term Evaluation of a Behavior-Based Method for Improving Safety Performance: A Meta-Analysis of 73 Interrupted Time Series Replications." *Safety Science.* 32: 1-18.

Krause, T.R., and R.J. McCorquodale. 1996. "Transitioning Away from Incentives." *Professional Safety.*

Lynch, P.D., R. Eisenberger, and S. Armeli. 1999. "Perceived Organizational Support: Inferior Versus Superior Performance by Wary Employees." *Journal of Applied Psychology.* 84: 467-483.

Manuele, F.A. 2001. *Innovations in Safety Management: Addressing Career Knowledge Needs.* New York: John Wiley & Sons.

Manuele, F.A. October 2000. "Behavioral Safety: Looking Beyond the Worker." *Occupational Hazards.* 62.

Manuele, F.A. July 1997. "Principles for the Practice of Safety." *Professional Safety.* 27.

Moorman, R.H., G.L. Blakely, and B.P. Niehoff. 1998. "Does Perceived Organizational Support Mediate the Relationship Between Procedural Justice and Organizational Citizenship Behavior." *Academy of Management Journal.* 41: 351-357.

Mount, M.K., M.R. Barrick, and G.L. Stewart. 1998. "Five-Factor Model of Personality and Performance in Jobs Involving Interpersonal Interaction." *Human Performance.* 11: 145-165.

Petersen, D. April 1998. "The Four Cs of Safety: Culture, Competency, Consequences & Continuous." *Professional Safety.* 32-34.

Roberto, M.A. 2002. "Lessons from Everest: The Interaction of Cognitive Bias, Psychological Safety and System Complexity." *California Management Review.* 45: 136-158.

Sulzer-Azaroff, B. and R.G. Mayer. 1991. *Behavior Analysis for Lasting Change, 2nd Edition.* Belmont: Wadsworth Publishing.

Tansky, J. W. and D.J. Cohen. 2001. "The Relationship Between Organizational Support, Employee Development, and Organizational Commitment: An Empirical Study." *Human Resource Development Quarterly.* 12: 285-300.

Vaidya, J. G., E. K. Gray, J. Haig, and D. Watson. 2002. "On the Temporal Stability of Personality: Evidence for Differential Stability and the Role of Life Experiences." *Journal of Personal and Social Psychology.* 83 (6): 1469-1484.

Wayne, S. J., R. C. Liden, M. L. Kraimer, and I. K. Graf. 1999. "The Role of Human Capital, Motivation, and Supervisor Sponsorship in Predicting Career Success." *Journal of Organizational Behavior.* 20 (5): 577-595.

Williams, S. 1999. "The Effects of Distributive and Procedural Justice on Performance." *The Journal of Psychology*. 133 (2): 183-193.

Zohar, D. February 2002. "The Effects of Leadership Dimensions, Safety Climate, and Assigned Priorities on Minor Injuries in Work Groups." *Journal of Organizational Behavior*. 75.

Zohar, D. 2000. "A Group-Level Model of Safety Climate: Testing the Effect of Group Climate on Micro-Accidents in Manufacturing Jobs." *Journal of Applied Psychology*. 85: 587-596.

Index

M

Machiavellian 116
Machine-guarding program 217
Management 124, 144, 153, 159, 199
  actions of 130
  definition of 133
  labor and 142
  lack of integrated 220
  support of 154
  task of 18
  trust in 233
Management by exception 39
Management chain 229
Management commitment
  link to safety outcomes 77
  meaning of 80
Management credibility 68, 71–72, 174,
  193, 195, 201, 216, 228
Management skills 229
Management sponsor 151, 203, 217
Management team 186, 192
Manager(s) 125, 216, 227, 233
  contact with hourly workforce 216
  maintaining an effective balance 230
  middle 116
  organizational goals and 18–19
  role in employee-driven process 200
  role of 154
  subordinate 234
Managing resources 209
Manufacturing facility 109
Manufacturing organization 1, 36
Manufacturing vice president 91
Marketing 2
Market cap 3
Market conditions 1
Maturity 35
Mayer, R.G. 90
Measures 169
Mechanisms and processes
  critical role of 20–22
  supervisors and 21
Medical errors 12
Meta-analysis 190
Metals and mining company 176–178
Military organization 174
Minority opinions 228
Misalignment 199

Mission Operations 233, 240
  at Johnson Space Center 246
Mission safety 224
Mississippi River 192
Mistrust 216
Misunderstandings 132
Money 37
Monthly reports 21
Montreal 199–200
Morale 1, 37
Motivation 95, 101
  of employees 37
  of leaders to improve safety 18–19
Motor vehicles 174
  fatalities 171
Mountain-climbing 106
Mountain Madness 105–106
Mount Everest disaster 105–107
Multi-level intervention 217
Multi-rater feedback 237, 239
Multiple objectives 137
Mutually supportive 214, 218
Mutual trust 133

N

NASA 21, 170, 219–249
Natural inclinations 43
  toward safety 19
Near miss(es) 150
Negative feedback 55
Negative issues 229
Night shifts 216
Nine dimensions of
  organizational functioning 193
Nitrogen 207
Non-enabled situations 130
Non-performance 239
Normative database 67
Norms database 178, 193
Northbrook 207
North America 213

O

O'Keefe, Sean 223
O'Neill, Paul 2–3
Objectives and vision statement 165–166,
  172, 177, 179, 183, 186

268

Index

DATE DUE

GAYLORD No. 2333   PRINTED IN U.S.A.